C000000291

797,885 Books

are available to read at

www.ForgottenBooks.com

Forgotten Books' App
Available for mobile, tablet & eReader

ISBN 978-1-332-41594-6
PIBN 10424299

This book is a reproduction of an important historical work. Forgotten Books uses
state-of-the-art technology to digitally reconstruct the work, preserving the original format
whilst repairing imperfections present in the aged copy. In rare cases, an imperfection in
the original, such as a blemish or missing page, may be replicated in our edition. We do,
however, repair the vast majority of imperfections successfully; any imperfections that
remain are intentionally left to preserve the state of such historical works.

Forgotten Books is a registered trademark of FB &c Ltd.
Copyright © 2017 FB &c Ltd.
FB &c Ltd, Dalton House, 60 Windsor Avenue, London, SW19 2RR.
Company number 08720141. Registered in England and Wales.

For support please visit www.forgottenbooks.com

1 MONTH OF
FREE
READING

at
www.ForgottenBooks.com

By purchasing this book you are eligible for one month membership to ForgottenBooks.com, giving you unlimited access to our entire collection of over 700,000 titles via our web site and mobile apps.

To claim your free month visit:
www.forgottenbooks.com/free424299

* Offer is valid for 45 days from date of purchase. Terms and conditions apply.

English
Français
Deutsche
Italiano
Español
Português

www.forgottenbooks.com

Mythology Photography **Fiction**
Fishing Christianity **Art** Cooking
Essays Buddhism Freemasonry
Medicine **Biology** Music **Ancient
Egypt** Evolution Carpentry Physics
Dance Geology **Mathematics** Fitness
Shakespeare **Folklore** Yoga Marketing
Confidence Immortality Biographies
Poetry **Psychology** Witchcraft
Electronics Chemistry History **Law**
Accounting **Philosophy** Anthropology
Alchemy Drama Quantum Mechanics
Atheism Sexual Health **Ancient History**
Entrepreneurship Languages Sport
Paleontology Needlework Islam
Metaphysics Investment Archaeology
Parenting Statistics Criminology
Motivational

"Eden"

An Excursion from New Orleans to the Pacific by Rail.

Through Texas & Mexico

VIA THE

"STAR & CRESCENT" AND "SUNSET"

STAR AND CRESCENT

AND

SUNSET ROUTE

"THE TRUE SOUTHERN PACIFIC."

GENERAL OFFICES, HOUSTON, TEXAS.

F 391
G19

Copyrighted, 1882,

By T. W. PEIRCE, Jr.

proprietor

AMERICAN BANK NOTE COMPANY, NEW YORK.

PREFATORY.

WITH the approach of cold weather, persons of delicate health, invalids and convalescents, wish to escape the long and rigorous winters of elsewhere, and feel that irresistible desire to go somewhere, which impels them to ask the all-important question: "Where shall we go?" It is the object of this little work to answer that interrogatory, and furnish the increasing number of tourists and seekers after health such information and suggestions as will enable them to find the desired haven in the winter resorts of this sunny clime.

Here are found the mild climate, the healthy atmosphere, the beautiful and romantic scenery, combined with a refined society and all the luxuries known to the enlightenment and civilization of the nineteenth century, in a degree not equaled by any resorts of similar character in the United States.

Without fear of successful contradiction, we can exclaim, "Eureka! The sanitarium for mental and bodily ailments has been discovered in western Texas." Thither you go, via the STAR AND CRESCENT and SUNSET ROUTE, so beautifully and happily named from the fact that it carries the tourist where the flitting sunbeams kiss their sweet good-night to Texas and to America. This line forms also a most important link in the great Southern Pacific, which, alone, reaches from ocean to ocean, and enables the busy denizens of both the Atlantic Seaboard and the Pacific Slope, to cross the continent from summer to summer, without the interruption of snow storms and ice blockades (often of weeks' duration) on the way. The managers of the STAR AND CRESCENT and SUNSET link are expending means and putting forth extraordinary efforts to add luxury to the comforts they already have for the ease and convenience of summer and winter travelers. Whether the trip is limited to a day or a month, it matters not, for the traveler can exercise his pleasure in stopping off at intervals, or continuing as his inclination dictates. Tickets for this especial purpose can be purchased of any and all ticket agents of the road.

From this time forward, passengers may travel either day or night, or both day and night, with the most perfect comfort, as the company runs two daily trains, which arrive at and leave the Union Depot in Houston, the great railroad centre of the state. On these trains are the elegant sleeping and parlor-car accommodations, which, in Texas, are to be found only on the SUNSET ROUTE, and which make traveling for the aged and infirm rather a relief than a tax. Oh, what a glorious change! What a wonderful contrast to the jolting of an old stage-coach, the sluggish drag of a "prairie schooner," or the exhausting jog of a broken-down horse; these, only a few years ago, were the sole methods of traversing this then wild and unexplored wilderness.

"THE TRUE SOUTHERN PACIFIC."

A MAGNIFICENT highway is now opened from New Orleans to San Francisco, and is naturally the nearest and most popular route to the West and Southwest. The tide of travel from European and Asiatic countries, and from the far-off islands of the great Pacific Ocean, can now cross this part of America, in less time and with greater safety, than by any other steam or railroad line on the globe. Industries will rise, and have already risen, by the opening of these wonderful trans-continental lines, and to two men chiefly is due the honor of this timely and great enterprise—C. P. HUNTINGTON and THOMAS W. PEIRCE. To them belongs the credit of having given to the world at large a new empire of trade, commerce, industry and agricultrue, by the construction of this grand highway in the interests of all nations!

STAR AND CRESCENT ROUTE.

THE crescent moon in oriental skies
Wanes as the stars of western empires rise;
And hangs above the ocean's eastern rim,
In pallid beauty, shining still, but dim.

HIGH upon azure fields of upper air
The glittering stars of Occident appear,
And blaze in noons of night and sink to rise
When sets the sun in occidental skies.

FROM desert lands of Araby and seas
Of tropic climes, and where the southern breeze
Fans languid leaves of citron, whose perfume
Makes drowsy atmosphere in isles of bloom,

REACHES an influence deep and ocean-wide,
Strong as the subtle might of rising tide,
Linking the eastern mount and western plain,
Binding the western shores and eastern main.

SCIENCE, thou goddess of the mighty West,
Lo! shining full upon thy generous breast,
And gleaming on thy brow; light seen afar!
Behold the mingled Crescent and the Star!

EMBLEMS of power, that embraces earth
From sea to sea in equatorial girth,
Mild as the moon's pale crescent and so bright
As ruddy stars in tropic skies of night.

OH, peaceful Union! Promise of a day
When different lights shall blend in single ray,
When earth shall move as one harmonious whole,
And mild-eyed Peace shall reign from pole to pole.

SPANISH FORT, ST. JOHN'S BAYOU, LA.

Amer'n B.N. Note Co N.Y.

From New Orleans to El Paso and Mexico,

VIA THE

"STAR AND CRESCENT" AND "SUNSET" ROUTE.

LOUISIANA.

THIS state, justly called the "Pearl of the South," has an area of 44,426 square miles; an area of 2,507,935 acres in cultivation; and a population (white and colored) of nearly one million within fifty-eight parishes (*i. e.* counties), of which Baton Rouge of East Baton Rouge Parish is the capital. Organized as the Territory of Orleans, March 3d, 1805, Louisiana was admitted into the Union under its present name April 8th, 1812. Seceding January 25th, 1861, it was readmitted into the Union, June 25th, 1868.

THE COMPLETED RAILROADS ARE:

Baton Rouge, Grosse Tete and Opelousas.
Chicago, St. Louis and New Orleans.
Clinton and Port Hudson.
Louisiana Western.
Morgan's Louisiana and Texas.
New Orleans and Mobile.
New Orleans Pacific.
Vicksburg, Shreveport and Pacific.
West Feliciana.
Texas Pacific.
Total miles of roads in state, nearly 800.

PROJECTED RAILROADS ARE:

Louisiana Midland.
Natchez, Texas and Pacific.
Vermillionville and Baton Rouge.
New Orleans and Northeastern.
Mississippi Valley.
New Orleans, Red River and Texas.
Total number of miles in and partly within the state, nearly 1000.

NAVIGABLE STREAMS.

Besides the railroads above mentioned, Louisiana has many water-ways, lakes, lagoons, bayous, rivers, etc., covering an area of over 3770 miles; and the U. S. engineers are now at work on a number of streams in Louisiana, and expect to add 240 miles of navigation to the water communication of the state within a year.

At present our space does not permit us to give more statistical details in regard to physical divisions, prices of United States and state lands of the state of Louisiana; but we may

speak of these in another issue. Nor can we dwell at length upon the history of Louisiana, but will commence our synoptical tourist's diary with the

HARBOR OF NEW ORLEANS.

In viewing the port of New Orleans, the stranger obtains on either land or river side so grand and eminent a sight that it is hardly possible for him ever to forget it. The levees of the metropolis of the Mississippi, as well as the Broadway of New York and the Chinese Quarter of San Francisco, may be considered as chiefest among the wonders of the New World. As an old traveler over North America for many years, I cannot remember to have ever seen, within so small a frame, such an enchanting picture as is presented by the port of New Orleans. In portraying the beauties of American scenery, one is too apt to become extravagant, but in this case it would not be out of the way to be so, for the sight is one that demands an extravagant admiration. New Orleans is the greatest *river* port in the world. London, New York and even St. Petersburg, cannot be called *river* ports, since, although they are situated on rivers, their ships are mostly anchored on the great waters. Here, however, you have a port on whose one side roll the grand waves of the Atlantic ; while on the other, the largest river on earth, together with its tributaries, opens to navigation a territory of a million English square miles. No other port can boast of that. New Orleans was destined to be one of the most prominent commercial points of the continent, and though many circumstances, as the civil war, political disturbance, and the rivalry of other places threatened at one time to dwarf its noble capacities, the city must become again the main port of the whole Mississippi valley, on account of its geographical situation alone, setting aside its other striking advantages. The last few years already show a steady and decided increase in commerce. From '79 to '80, one-third of the whole cotton crop of the United States was conveyed to New Orleans, while New York received but one-fifth. The metropolis of Louisiana is the chief cotton centre of the Union, and this can be easily believed after a sight of the river at that city. The Mississippi here is not nearly so wide as at other places, but it has still a width of a kilometer, and a depth varying from 150 to 200 feet, preventing all danger of ships foundering on sand banks. The docks of the steamers and sea vessels are on the left side of the river, and upon the opposite side are the steamers and ferryboats of the Morgan line (that has been of so great assistance in the development of the commerce of New Orleans), and the depot of the but lately finished New Orleans and Texas Railway. In the midst of the river one sees a few ocean steamers, three-masted vessels and brigantines, and at the upper part of the port, lying at anchor, is a heavy American war-monitor, like the grim gate-keeper of the City of the Half Moon. These vessels, hardly moved by the high waves of the river, are only in the background of the scene that opens for miles on the long levees of the left side of the river. Dry-docks, harbor-basins, stone-quays and the inevitable stereotyped necessities of all seaports, are quite unknown here. Their place is taken in New Orleans by a parquet, floored and made of heavy logs, extending along the whole river, partly resting on the mainland and partly floating on the water, the whole firmly fastened by anchors. At a probable estimate, this gigantic raft has a width of about 50 paces. Behind this there is a wide unpaved place stretching for miles along the river, which is the terminus of no less than 150 streets. Railroad tracks cross the shore in all directions. Depots, warehouses, canvas sheds of immense extension, and finally, mountains of cotton bales, sugar and petroleum barrels, logs and lumber, occupy, here and there, parts of this extensive locality, and are far from filling it. Railroad trains, dozens of street cars, and hundreds of vehicles of every description, loaded or unloaded, pass up and down the streets in long uninterrupted trains. Thousands of people —office-holders, merchants, sailors, policemen, travelers, dealers in fruits and vegetables, and last, though by no means of the less importance, dock-hands—pass swiftly and busily among the cotton bales, agricultural implements and other articles peculiar to docks and wharves.

NEW ORLEANS THE METROPOLIS OF THE SOUTH.

We quote here a pen sketch by the able writer, John E. Land :

"Rich land! Noble history! A land so fertile, God seems to have pronounced upon it his sweetest benediction. A climate so mildly tempered, 'the mock-bird has no winter in his song, no sorrow in his year.' A soil so generous, it gave ample competence to all who came, and afforded ability to indulge, not only in those pursuits which tended to satisfy animal wants and desires, but softened into poetry the selfish passions, improved the moral and intellectual character and gave leisure for liberal studies and pursuits. Thus, with that tranquility and leisure afforded by the enjoyment of accumulated riches, those speculative and elegant studies which enlarge views, purify tastes, and lift mankind higher in the scale of being were successfully prosecuted ; and thus do we account for the illustrious names New Orleans has furnished to the world—in law, in medicine, in divinity, in judicature, in commerce, in military science and literary accomplishments—names that enrich not only the biographical wealth of the city, but have been enrolled among the nomina clara of the Republic.

"It should be an accepted fact, therefore, that the barbarism or refinement of a people, whether national or municipal, depends more on their wealth than on any other circumstance. No people have ever made any distinguished figure in philosophy or the fine arts without being celebrated at the same time for their employed riches and industries. Pericles and Phidias, Petrarch and Raphael, adorned the flourishing ages of Grecian and Italian commerce. The influence of productive wealth in this respect is almost omnipotent. It raised Venice from the bosom of the deep, and made the desert and sandy islands on which she is built the powerful 'Queen of the Adriatic ;' it rendered the unhealthy swamps of Holland the favored abodes of literature, science and art ; and it has done as much, and will do vastly more, for New Orleans, the Imperial City of the Gulf—the midway mart of North and South America.

"While it is true there are a few, even of our best informed citizens, who are sceptical as to the continuance of this magnificent prosperity, and are evermore on the lookout for sudden and fatal checks to the city's trade and enlargement, it is equally as true that they fail to notice fully either what has already been

BAYOU TECHE.

accomplished or the unlimited resources about us yet undeveloped, but certainly to be drawn upon, in the grander conquests of the not distant future. Let us, therefore, be candid with all such, and assert without fear of successful contradiction, that the very best assurance of the continued healthful progress of New Orleans is found in what she is to-day—a centre of enormous trade, in spite of some of the most unfavorable surroundings and drawbacks that ever beset a city, more, perhaps, the creature of the necessities, the inexorable demands of the position, than any American city that has ever struggled for eminence ; and yet the forces that have thus successfully built up the city are far from being exhausted, or even fully comprehended. Humanly speaking, then, there is no power on earth that can prevent New Orleans from becoming a vast commercial city. It will grow in wealth and power, in industry and influence, in spite of itself—in spite even of the bad fame she has abroad on account of climate. The demands of commerce, like the demands of necessity, know no law, admit no obstacles, overcome all barriers. Back of the city to north, to east, to west, lies a vast empire of productive wealth with many millions of people, all of whom, in a manner, are ministering to its traffic and wealth. Like fabled Cerberus, who guarded the entrance to Pluto's realms, New Orleans mounts guard on the highway of the Mississippi valley ; and whosoever approaches will be challenged, and whosoever passes must pay tribute for the privilege of egress or ingress through this grand gateway, this unrivaled outlet, this natural inlet of trade and travel from the heart of the American continent to every land and clime and sea where the flag of commerce is unfurled.

"In truth, the view is propitious from every standpoint. The city is in a condition of vastly improved sanity and health, and has commenced—nay, is far upon the road in a brilliant career of improvement. The motives of social and political freedom, fertility of soil, salubrity of climate, wealth of agricultural resources, facilities for commerce and manufactures, and ease of river and railroad transportation, are the material advantages which invite capitalists, tradesmen and manufacturers of every clime and nationality to a home in our midst, to a cooperation in the development of its measureless possibilities, and to an enriching participation in its prosperity. A live, intelligent and enterprising people, now fully aroused to all the requirements of the age, have possession of her multifarious labors, and the day is now at hand when many a stately edifice is musical with clanging machinery and those sounds of diversified industry that quicken the pulse of a nation and prolong the life of a republic ; while her possibilities, thus foreshadowed, dazzle the mind by their variety and magnitude, and leave the calmest and most unimpassioned observer quite bewildered in the prospect for this magnificent metropolis of the New World."

THE "CRESCENT CITY"—WHY SO CALLED.

The older portion of the city is built on the convex side of a bend of the river, which here sweeps around in a northeast, east and southeast course. From this location it derives its familiar sobriquet of the "Crescent City." Whether we take it in the garish light of day, or under moonlight or starlight vision, no city of the New World presents a fairer view than New Orleans, the Crescent City of the South. Whoever has seen its multiplied charms by day will pardon the enthusiasm of the writer who described its charms at night and as seen from the deck of one of our noble steamers : "The mantle of night ·has settled on the scene, and the historic Crescent City, with a myriad of gas-jets beaming, seemed, as we approached, a picture from fairy-land, instead of a reality. Quite romantic and bewildering is the view as we round the bend and come onward down the stream ; and it did seem that the rolling flood of the Patriarch of Waters had merely made this graceful curve, as if it longed to look upon a spot of so much beauty ere it journeyed on in its unceasing travel to the remorseless sea. Bending like the curve of a Mussulman's cimeter, with each light from the shore reflected from its bosom, the sight was indeed Oriental and crescent-like ; and one might easily add in imagination the crescent-standard battalions of the Grand Sultan, and picture the hosts of Islam passing in view. Yet, by its shape alone, does our beloved city claim the symbolic name of the Mohammedan, and we owe no obeisance to Saracenic poetry for the suggestion.

"However, in the progress of its growth up stream, the city has of late years so extended itself as to fill the hollow of a curve in the opposite direction, so that the river front now presents an outline somewhat resembling two conjoined crescents, or perhaps more properly the letter S. This configuration necessarily renders the direction of the streets very irregular."

The old Spanish Fort, which over a century ago was the stronghold of the early colonists was erected to protect the mouth of St. John's Bayou. Following the windings of the stream, at that period, resulted in the discovery of the present site of New Orleans. What marvelous changes ! Nothing left of the old fort but some crumbling walls, at present ornamented with huge live-oaks. Wandering over these historic grounds, you discover as relics of a past time two dismounted ancient cannon, half buried in the grass ; and on the side of the vanished fort, among the trees, now stands a summer residence, enlarged to accommodate visitors of the newly laid-out grounds. We are informed that some years ago it was a favorite summer resort, with rose gardens, orchards and orange groves bright with the golden fruit of Hispania. Sailing vessels plied on the bayou ready to take the delicious fruit to distant marts. This orange grove was totally destroyed by the great flood a few years ago ;

the estate changed hands several times, until the railroad company took charge of it and made it the most attractive place of the kind in the South. Ascending the great pavilion, eighty feet high, just before sunset, a beautiful scene is beneath you. You can trace the course of the river far below and see the levees crowded with steamers and a motley array of vessels. Nearer, you can see the city over the intervening swamps, with its many steeples and stately buildings. Turning toward Lake Pontchartrain, you see its waters until they unite with the waves of the Mexican Gulf. Looking eastward you can see the lights of West End, with its numerous buildings—a rival of Spanish Fort. It was erected shortly after the Spanish Fort enterprise by another railroad company. Further on is the bayou, covered with many canoes and sailboats gliding as smoothly as Venetian gondolas beneath the soft radiance of the southern moon. What Italian landscape could be more beautiful?

THE CRESCENT CITY DIFFERS FROM OTHER CITIES IN AMERICA.

New Orleans is a city of surprises. The visitor, during even a brief stay in this remarkable place, is struck with the greatest contrasts in merely walking from one street to another. He will pass through thoroughfares in no way inferior to those of Paris, Vienna, St. Petersburg, Palermo, Venice or Rome. He will find the nervous and restless business life of the Yankee in contrast with the "dolce far niente" of the Creole; he will see many interesting pictures in the lives of the half-civilized negroes, or in the scanty existence of a few Indians; he will meet representatives of every nation in the world, and hear languages and dialects of every country. He will soon discover that he is in a city demanding a long stay and diligent study in order to a clear comprehension of her and her wonders. Some European tourist once remarked sarcastically, that having seen *one* American city he had seen *all;* but, without doubt, he had never been fortunate enough to visit New Orleans or San Francisco. A lengthy visit to either of those places would speedily disabuse his mind of the error into which he had fallen, unless he is a victim to old world prejudice.

A VISIT TO THE FRENCH MARKET.

The French Market before sunrise is an ideal scene of lively traffic particularly interesting to the traveler from the North, who has never seen anything like it at home. During the morning hours of each day these markets are veritable beehives of industry. Pretty young negresses chat with monsieur and madame, who by the expenditure of only a few cents can get a day's supply from the generous quantities of tropical fruits, such as bananas, oranges, pineapples, etc. Mulatto belles, who call themselves Creoles, may be seen gracefully gliding through the crowd—and, in fact, the stranger who does not know the difference between an octoroon and a Creole might easily be mistaken, for they imitate Creole manners to perfection; ladies with their servants, children of every color, old men and dandies, flutter along this endless line of exhibited merchandise, between two of the great market buildings, to buy either vegetables, fish, fruit, meat or toilet and fancy articles. Every vender is in search of a customer, and tries to persuade passers-by to bargain with him. On all sides rises a great noise of voices chattering in a dozen different patois — French, English, Spanish, and the guttural negro dialects. Occasionally one of those superb, almost faultlessly formed Creole ladies, passes through the throng unescorted; she makes her scanty purchases with the same grace as in former times, when wealthy, she stepped from her carriage to enter one of the magnificent jewelry establishments to purchase precious stones as negligently as she now pauses to buy fresh fruit or fish.

One notices on every hand a profusion of flowers, bouquets of freshly culled roses, fragrant orange blossoms, delicate jessamines, waxen japonicas and camellias, purple and gold and gorgeously hued tropical blossoms— for Flora is the divinity of this balmy clime. These flowers are truly marvelous, and we believe that no other city in the world has daily such lavish and wonderful display; and the most unusual and rare blooms are sold at so small a sum that one can scarcely credit it. Before the intense heat of noon wilts the fruit, flowers and vegetables, the jargon of languages dies away, the endless procession fades like phantasmagoria; the stalls and booths are silent and empty, and the golden sunlight streams through the deserted market-place.

TOPOGRAPHY OF NEW ORLEANS.

New Orleans, unlike the majority of American cities, is not laid out in squares, and this circumstance alone would prove it to have been built by the French. Although a few of the streets are straight and some parts of the city are built with strict regularity, many of the thoroughfares are parallel with the bends of the river or with the once great private estates. The oldest and real business part of the city has been built into a curve of the Mississippi, while the newer parts lie in the direction of Lake Pontchartrain. A few bayous and canals traverse the city in various directions, but the "Grand Canal" of New Orleans is the Mississippi River, the father of waters. On it alone depends the flourishing condition and the future of the metropolis, the "to be or not to be" of the Crescent City. Those streets next to the river-port are in the centre of the business life; but if you follow their windings, you will find yourself in the quieter portions of the city, and will finally be lost among private residences, alleys and gardens. A number of straight avenues cross the squares for miles in various directions, converge and diverge, and lead in some places to the river or to the lake, without ever having a real centre. Of the multiplicity of streets not a dozen run in the same direction. It would have been impossible, and without any practical benefit, to have numbered them, as is often done in large American cities ; accordingly they have, in New Orleans, particular names, as in European cities, and these names have a manifold origin. In the older parts of the city, for instance, the names recall the memory of the French kingdom ; some of the later ones, the Republic and the Empire, as Austerlitz and Marengo Streets. Other appellations belong to the Muses, to demi-gods, fairies and Titans. In short, ancient mythology was raided in order to supply names for these modern American avenues. The "Path of Fortune" crosses the Dryades, Elysee, and Magnolien Streets, and finally the "Path of Melpomene." Columbus finds himself in the society of charming demi-goddesses, and the French kings are surrounded by the highest gods of Olympus. Can one go further in a kingdom in adoring royalty ? Our national heroes and presidents were carried into the neighborhood of all virtues, as Genius, Wisdom, Power, Charity, etc. On the other side of the stream are the suburbs of Algiers, Tunis and Belleville. A plan of the Crescent City must look, without doubt, much nobler with such flattering and ostentatious and poetical names, than plans of other American cities with their plain numbers 1, 2, 3, etc.

CANAL STREET.

As space will not permit a more detailed description of New Orleans, we will at least accompany the reader into the most important streets. Starting from the Mississippi levee, we find ourselves in Canal Street, the principal and most beautiful street of the Crescent City, despite its less poetical name. It is the worthy associate of the great river to whose banks it leads ; it is the Mississippi of the streets. In New York it is Broadway that shows us the whole continent in the space of a few miles, a street on whose one end is Europe, and on whose other end is America. What Broadway is to New York, Canal Street is to New Orleans, alike impressive, alike grand, and perhaps even more characteristic. We see that as soon as we enter the street. It is the main artery between the South of the Union and the Tropic of the West Indies. The manner in which it is laid out, its appearance, the crowds of carriages and pedestrians, the immense and showy shops and stores, prove it to be the meeting point of the Anglo-Saxon and Franco-Spanish culture, as well as the line of meeting between the temperate and tropical zones. .These physical and mental contrasts may be found in this one city and more especially in this one street. New O-leans is divided into two almost equal portions by Canal Street, which has a width of about seventy feet. Lined on both sides by magnificent buildings whose height, number of stories and magazines, show the American energy, it has also structures whose architecture, breezy balconies and romantic verandas display the love of the Latin race for ease and comfort. Flowers in the windows, tropical plants on the balconies—a picture of the Spanish or French South ! Large advertisements, signs painted in glaring colors, windows gay with costly exhibitions—a picture of the American North. The names which we read on the street, the languages in which they are written, give another example of the Franco-Spanish American Babylon. Long flagstaffs, with stars and stripes, reach high over the houses. Green shutters and shady roofs of canvas are in the balconies ; in the street windows various sights, now an advertisement, now the face of a business man, and farther on the dark eyes of a charming Creole beauty. This long row of buildings for miles is only interrupted by other streets, whose terminus is Canal Street. The sidewalks, covered with great paving stones, are very wide. In the midst of the street, which is somewhat lower than the sidewalks, but very well paved, there are long rows of shade-trees, and footpaths well filled up with sand. In this way the space for vehicles is divided into two streets, partly separated from the sidewalk and shady footpaths, by small open ditches, which are covered only at the crossings. These secondary streets, still twenty paces in width, are crossed by dozens of tracks, and the travel on them, by horse and steam-cars, especially at the crossing from Charles Street to Canal Street, is enormous. Canal Street is in the centre of the net of street railways, which have an extension of over a hundred miles, and which are of great importance on account of the great distances. The street-cars are drawn by mules, as they can better endure the extreme heat in the summer than horses. We might say something of Carrollton Avenue and the Esplanade, both very handsome streets, used as promenades and drives by the fashionable and wealthy inhabitants of the city, but instead we will take a glance at Jackson Square, formerly the "Place d'Armes" of the French city. It lies in the midst of Frenchtown, and is one of the most attractive squares in the Crescent City, with its tropical verdure and tall orange and magnolia trees, whose white blossoms fill the air with heavy sweetness. Besides these, bananas, shrubbery and flowers of all sorts abound. In the centre of the garden is the equestrian statue of General Jackson, farther on is the Cathedral of St. Louis, with its three tall spires, that can be seen long before you reach New Orleans. On both sides of the Cathedral are two old buildings, used now as court-rooms. The park-like garden leads toward the Mississippi, and the entire square stands in marked contrast with the ancient and homely buildings around it in old Frenchtown.

Another attractive feature of New Orleans is the City Park. This would be one of the finest parks in America, for nature has done much in ornamenting the grounds with majestic live-oaks, festooned with gray Spanish moss. At present the park is not opened to visitors ; it has been fenced in, and is rather a cow-pasture than anything else. Near the park is a bayou which would make a beautiful lakelet if properly cleaned, but the people do not patronize these shady retreats as they should do, and it is nothing but a home for cows and geese.

Among the various societies of New Orleans is the new Louisiana Jockey Club, with about 300 members. This club has opened a fine race-course for the advancement and improvement of racing and horse-breeding. One of the loveliest parks, the property of the club, has been supplied with all the necessary buildings for these purposes. Here are given delightful promenade concerts, to which members only have admission. An invitation is a compliment extended only to visitors, and the invited stranger may consider himself a lucky man, as the elite of the Crescent City society assembles there.

Another society is the Liedertafel, with over 700 members, the largest private club in the city. The members of this society give musical and social entertainments, and during the carnival season they take a prominent part when the mystic Rex issues his proclamation. They have fine club-rooms and elegantly furnished drawing-rooms, in fact every accommodation for the lady and gentlemen members of the club. The New Orleans Philharmonic Society has about 100 members. The club is not only a literary and musical one, but also aims to improve the public taste in art. There are elegant rooms for concert and art purposes; but Grunewald Hall is used for chorus and oratorio singing. This hall is an ornament to the city, being of fine architectural proportions and having its acoustics as nearly perfect as is possible. The ceiling is beautifully frescoed and the walls adorned with many valuable paintings. The Philharmonic Society not only make use of the building, but other musical associations, comprising the finest talent of the city, give entertainments here, and on the evening of Mardi Gras an immense and loyal crowd assembles to pay due respect to Rex and his queen.

In this connection we may mention the "Carnival Club." The Variety Theatre presents, during the progress of the Mardi Gras ball, one of the loveliest sights in the Crescent City. From floor to ceiling, the parquet, dress-circle and galleries are one mass of dazzling toilets, for none but ladies are given seats. White robes, delicate faces, dark sparkling eyes, luxurious folds of glossy hair, tiny, faultlessly gloved hands—such is the vision that a humble spectator of the masenline gender may see through his opera-glass. Delightful music swells on the flower-fragrant air ; the tableaux change like pictures seen in a kaleidoscope ; the curtain rises, and joyous grotesque maskers appear upon the ball-room floor ; gradually the ladies and their cavaliers leave all parts of the theatre and join the dance, and the mystic Rex holds his levee.

The Yacht Club have a fine pavilion at West End, on Lake Pontchartrain, and have rules and regulations like the Jockey Club. The club regattas are celebrated. Another association similar to this is the "St. John Rowing Club," that gives festivals with fireworks on the lake. Other societies, as rifle clubs, turners, secret or charitable associations, abound.

In regard to public buildings, both finished and unfinished, New Orleans is second to no other city. We might mention the Cotton Exchange (the new building), Grunewald Hall, Lilienthal's Art Rooms, St. Louis Hotel, the Supreme Court House, the United States Branch Mint, the Custom House with the Post Office, Jewish Synagogue on Lee Place, St. Paul's Church, Trinity, St. Joseph's, St. Patrick's, Jesuit Church and College, Christ Church, St. Anna's Asylum, Odd

Fellows' Hall, Marine Hospital, Maison de Sante, the University of Louisiana and the noblest of all institutions—the Charity Hospital. To give a description of prominent business houses, or private villas and residences with their fairy-like gardens, would make an interesting volume in itself. But taking leave of the Pearl of the South, and casting a lingering look of regret at this lovely spot, we cross the Mississippi on the Morgan ferry and land in Algiers, the depot of the Star and Crescent Route. We soon obtain seats in the smoking parlor-car, and are off to visit some sugar-cane plantations, and to note down all that is worth seeing from New Orleans to the city of Houston. While comfortably smoking in the coach we make the acquaintance of a Creole planter, whose ancestors, he tells us, were among the first colonists. What a change in one hundred and fifty years! New Orleans was then the capital of an empire-colony that comprised a territory of over 500 miles along the Pacific, 1700 miles along the British domains, about 1400 miles along "Meschashebi," the Father of Waters, and over 700 miles along the Mexican Gulf. Louisiana at that time was the Western dream of every European crown. Two Indian tribes, the Choctaws and the Chickasaws, were in a state of constant warfare, and the colonists had to struggle for a precarious existence under circumstances of extremest difficulty. Out of this immense territory the finest states were formed, and the name Louisiana, belonging originally to an immense tract of one and a half million square miles, was transferred to a territory of only 40,000 square miles. Louisiana has a great future revenue in her cane-sugar production. There are in the state from twelve to thirteen hundred sugar-houses, of which nearly one thousand are run by steam. It has been stated that the sugar district of Louisiana is about 13,000 square miles in area, divided into nineteen or twenty parishes and showing the largest population in the Mississippi delta. Although Northern capitalists have taken possession of many old plantations, whose new chimneys and clanking machinery destroy the ideal landscapes, yet there are many happy homes still left in their old-time picturesqueness along Bayou Teche, Bayou Lafourche, Bayou Sarah, Baton Rouge, Natshita and other places up and down the Mississippi, where Creoles and old French settlers "hold in mortmain still their old estates."

Another important part of the state is the Red River region. Red River enters Louisiana at Shreveport, traverses the state diagonally and flows between Natchez and Sarah into the Mississippi. This part of Louisiana has at present too much water, and is but thinly settled. Still cattle in large numbers, various products and lumber are brought by steamers to New Orleans, and as we cannot give a detailed account of this section, we will mention that only six parishes along Red River are adapted for sugar-cane and cotton cultivation, while the interior region produces grain of every kind. About 9000 square miles wait for the agriculturist to open new avenues of wealth. Altogether, the southwest of this state has yet several millions of acres of excellent soil lying untilled. Toward the gulf, along the rivers, bayous, lakes, lagoons, are gigantic cypresses and other forest trees waiting for the woodman's axe, and between the parishes of St. Mary and Iberia, Vermillion, St. Martin and Lafayette (the former home of the Attacapas), are at present the best cultivated plantations of Louisiana, girdled by a belt several miles wide, containing majestic forests with all kinds of useful woods peculiar to the South. Farther down, along the gulf, stretch the savannas, the home of uncounted cattle; then come prairies yet untouched by civilization and but little known, but which will soon be opened by enterprising railroad companies. A brief sketch of the towns and plantations traversed by the Star and Crescent Line to Houston will perhaps be interesting.

Jefferson Parish has an area of 395 square miles, with 19,767 acres under cultivation. Population, 12,166. Seat of justice, Gretna, an old place with evergreen orange groves, stately live-oaks, fine gardens and ancient houses. In the background is the river and the levee, lined with dense woods. The Company Canal connects at Jefferson with Grand Isle, a watering place. A little steamer makes regular trips during the time from May to October, for the accommodation of passengers, and to carry freight. Toward Junction are seen fine sugar-cane fields in good condition; now and then one catches a pretty view of sugar mills and adjoining buildings through pleasant groves. We notice among various shrubs along the road, a "fleur du sureau." The white blossoms in clusters are beautiful, and the berries make a delicate confection, and also wine and vinegar, of which the Creoles are very fond.

Nineteen-Miles-Switch is a small place with rice fields in cultivation everywhere. Looking over the country one sees the pale green sugar fields and waving plains of rice, separated by hedges; forests, looking hazy-blue in the distance, and chimneys of sugar-mills on fine plantations near at hand. St. Charles Parish is next, with an area of 284 square miles; 21,177 acres in cultivation, and a population of 7,161. The seat of justice is Hahnville. St. Charles and Boutte, twenty-four miles apart, are passed, and we catch glimpses of cottages, gardens, and busy workers in ripening fields. A vast amount of land is still uncultivated. The soil of the parish is good, and the bottom-land is covered with dense woods, containing a great variety of shrubs and trees, festooned with superb vines. The tourist will be delighted with the lovely scenes. Twenty-eight-Miles-Switch is a water-station, and is a pretty stopping place, with its clear lake covered with snowy lilies, and overhung by moss-covered branches of the live-oak. The scenery between Boutte and Mestier is of the same character and its charm increases as we reach Bayou des Allemandes, in the middle of which is a tiny island upon which is a picturesque old cabin, hidden by trees of richest foliage. The stream is full of cypress logs, which are shipped from here to the gulf. Between green prairies rimmed with forests, is an opening toward the gulf, the verdure contrasting charmingly with the blue skies above. Cane-fields, dark forests, low shrubs, prairie land, flash past in dazzling variety. Raceland, forty miles on, is only a single cottage and water-tank. Again come rice, cane and corn fields, mills, houses and hedges on both sides; sufficient proof, it would seem, of the salubrity of the climate and the fertility of the soil. The only danger to be apprehended here is the occasional inundations, but which will be prevented in future by proper river engineering. At Ewing's, where we met the Houston train, we saw a monster alligator swimming in the water. He was soon a target for the revolvers of the passengers, but with the imperturbable indifference characteristic of his species, he swam placidly along, and soon was out of reach. The country here is more elevated, and the plantations show excellent culture. No doubt from the indications, a rich harvest of sweet potatoes will be reaped here. If the country were properly drained, the soil would be unusually fertile, and fruits of all kinds would grow in abundance. Lafourche Parish has an area of 1024 square miles; acres in cultivation, 44,802; population, 19,113. Thibodaux is the county seat. Lafourche is the next station, fifty-two miles, situated on a bayou; it is a quiet town. The bayou has very clear water, is navigable, and is a tributary of the Mississippi. The levees of this canal are from eight to ten feet high and extend to the river; there is also connection by lagoons with Morgan City. Between Lafourche and Terrebonne, a distance of fifty-five miles, all the good land is in cultivation, interspersed with groves of fine trees. The Land Reclamation Company is digging canals to drain this section, and doubtless much valuable land will be gained. The bottom-land along the streams is considered to be some of the best in the whole state. Terrebonne Parish has an area of 1806 square miles; 40,403 acres under cultivation, and a population of 17,956. Houma is the seat of justice. Towards Chacahoula, sixty-one miles, we can see cypress and water-oak swamps, the resort of alligators, turtles and water-snakes. Snow-white cranes stand silently on aquatic plants waiting for their prey, and beautiful palmettoes, from ten to twenty-five feet in height, spread their graceful leaves in the damp shade. We find a similar landscape en route to Tigerville, L'Ourse, and Boeuf Bayou, connecting with Atchafalaya River, near Morgan City. This is a place where lumber of all kinds is shipped to different points of the country. At Romos some plantations are in sight. Years ago the Acadian exiles came to this region from Acadia, cut down the cypress forests and, following this unnavigable portion of the Bayou Teche, settled near its source and began to cultivate the ground. Many comfortable houses are seen hidden in groves of Pride of India trees, built in plain cottage style, weather-boards outside and plastered walls within; in front the inevitable veranda. The corrupted French of these Acadians is the same that was spoken years ago. They marry at an early age, and it is stated that girls twelve years of age, and boys a few years older, become husband and wife and take charge of their own homes. All these settlements reach far beyond the prairies of Opelousas. To be sure, these people live in the United States, but as to being Americans—that is quite another thing! In these regions many lagoons, connecting all these waterways, are known only to these Acadians. For hundreds of miles swamps, with all their mysterious contents, wait to be traversed by man. And what will be the future home

of industry is now the retreat of the moccasin and alligator. From here towards the gulf and Mississippi, a distance of over 150 miles, there is nothing but swamps. The country around Romos Bayou, a branch of Boeuf Bayou, is mostly inhabited by negroes. The fields are in good condition and bring good crops. At Morgan City, eighty miles further on, we cross, on a fine iron bridge, the Atchafalaya River to Berwick Bay, a place celebrated for its excellent oysters. Boeuf Bayou and Tedre River are tributaries of this bay, and a live trade gives Morgan City the importance of a large town.

After passing Pattersonville, farm joins farm, and here flows the beautiful Teche (corrupted from *Deutsche*) for sixty miles, its lovely shores in gentle curves. Bartles' Plantation is a little

town in itself, with its fine buildings, shady groves, sugar mills, orchards and good-looking negroes and negresses. In the fields, not far away, are the "quarters," rows of cottages, each with a veranda, and all shaded by fine trees. St. Mary's Parish comes next; seat of justice, Franklin; area, 648 square miles; acres under cultivation, 66,326; population, 19,891. Bayou Sali shows beautiful forests; and the next place we reach is Franklin, a lovely town of 2800 inhabitants. Farther on are Baldwin, Sorrel, Jeanerette and New Iberia. It is as if we see nothing but one great garden. There are fine residences, plantation after plantation, bosky woodlands, rolling prairies dotted with fine stock, corn and sugar in the highest state of cultivation. The most beautiful portion of fair Louisiana lies along Bayou Teche. Nature discloses here all her charms; and taking passage on the steamer that plies on the clear water, you are gliding in "the lost Eden," an earthly paradise, where life is good because of the delicious investing of it by nature with everything that is fairest. When you wish to see plantations at the height of culture—lawns as fragrant, as clean-shaven, as nobly shaded by graceful trees as any sovereign's, then seek the Teche country ; it is the gem of Louisiana ; it is the perfection of the South. Thither Andry and the exiled Acadians took their mournful way more than a century ago, when the cruel order of the arrogant English dispersed them from their homes ; thither they went, threading the swamps and wandering up the beautiful Atchafalaya and her lakes, where

> " Water lilies in myriads rocked on the slight undulations
> Made by the passing oars, and resplendent in beauty the lotus
> Lifted her golden crown above the heads of the boatmen.
> Faint was the air with the odorous breath of magnolia blossoms,
> And with the heat of noon ; and numberless sylvan islands,
> Fragrant and thickly embowered with blossoming hedges of roses.
> Near to whose shores they glided along, invited to slumber."

Now, as then, the traveler pushing his way in a tiny steamer, or in a shallop or piroque, can hear

> "Far off, indistinct, as of wave or wind in the forest,
> Mixed with the whoop of the crane and the roar of the grim alligator,"

strange sounds from the dark forests and the lonely lands.

From Berwick's Bay, where the rich fields lie trustingly upon the water, strange vines and creepers seeming to caress the waves and bid them to be tranquil, ascend the Tedre Bayou and lose yourself in the tangled network of lake and lakelet, plain and forest, plantation and swamp. By day you shall have the exquisite glory of the sun, which, gleaming on the seignorial residences, the great white sugar-houses with their tall chimneys, the long rows of cabins for the laborers, the villas peering from orange groves—makes all doubly bright and beautiful ; and at evening the moon may lend her witchery to swell your admiration and surprise. You will drift on by superb knots of shrubbery, from which the birds sing amorous madrigals ; past floating bridges and garden bowers ; along the banks by a ruined plantation, one of the wrecks of the war. Now seeing in the distance dense cypress swamps, bordered by picturesque groupings of oak and ash and gum trees; now through that fine region extending from the entrance of the bayou into the parish of Iberia and the town of New Iberia, where the beautiful water-willows and forest trees lean downward from the banks to see themselves reflected in the stream, and where the wheels of passing steamers are compelled to brush them rudely as they pass between ; where the live-oak spreads its ample foliage over some cool dell, upon whose grassy carpet strange bright-hued flowers grow rankly ; and where suddenly, as though opened by the hand of enchantment, a vista of forest glade, of happy sylvan retreats, where the moonlight makes checker-work of gleam and shadow, appears before you.

New Iberia is the county seat of the parish of Iberia. Area of the parish, 536 square miles ; acres in cultivation, 49,664; population, 16,686. From New Iberia, a distance of 125 miles on our way, you can visit the romantic and untouched forests, full of game, along the shores of the ".Des Lac Peigneur," or go to the well-known Orange Island. At New Iberia some enterprising gentlemen have done much for horse breeding—something very much needed in Louisiana. Near Cherondon, a village about six miles from Sorrel, the Teche makes a great bend, and this whole part—including Jeannerette, a lovely town of about 1500 inhabitants—to Iberia and farther on, is the wealthiest part of Louisiana, of which Oliver Plantation, with over 1000 acres under fine cultivation, is considered to be a model. Mr. Oliver is an authority on sugar and fruits in general. " Below New Iberia, on Petit Anse Island, you may descend into a salt mine sixty feet beneath the level of the Gulf of Mexico, through fifty-eight feet of solid rock-salt, and watch the miners picking out the crystal freight, which has proved superior to any other salt found in the southern market." Toward New Iberia, which is on the line of the railroad, fine prairies, with excellent cattle feeding on the nutritious grass, alternate with highly cultivated cotton fields ; sugar-cane fields alternate with gardens for miles up to Vermillionville. New Iberia has a population of nearly 4000 inhabitants ; has cotton-seed mills, sash and saw mills, foundries, shingle mills and a large trade in lumber. Railroad tracks have been laid to the different saw-mills, for greater convenience in carrying off the logs, which are generally cypress. There are two good schools here, one Catholic and the other Presbyterian. About six miles farther on we come to St. Martin, on Bayou Teche ; the town is a chief shipping point, and is situated on a branch of the railroad. We come next to the lovely Spanish Lake, with its fine residences, the water as clear as crystal and full of fish and water-fowl. The lake is over three miles wide, and is seven miles long. Along the shores are excellent orange groves, the moisture protecting them from frost. The landscape is lovely, and ash, pecan, live-oak, wild cherry, water-oak and cypress make beautifully shaded groups of green. The climate is very healthy from Franclin to Estherwood, and fevers are unknown. Toward Broussardville and Royville undulating prairies predominate. The country has natural drainage, and corn, cotton, sugar-cane, fruits and all sorts of berries grow here luxuriantly. Almost anything will grow here, and the Acadians must have known that, and therefore chose it as a settling place. The view at sunset is grand, and Spanish Lake looks like molten gold. St. Martin's Parish has 648 square miles, of which 39,876 acres are under cultivation. Lafayette Parish has an area of 262 square miles, of which 62,704 acres are being tilled. The population is 13,236. Between Rayne, 159 miles from New Orleans, and Vermillionville, the county seat of

SOUR LAKE—HARDIN COUNTY.

Lafayette Parish, everything shows the industrious character of the people. Rayne is quite a town, has good stores and neat cottages. You see farms on all sides with well stocked cattle-pens and very fine horses; cattle ponds supply the stock with an abundance of good water. The soil is chocolate in color, and well adapted to the raising of sweet potatoes. St. Landry Parish has an area of 2276 square miles, and an area of 137,370 acres under cultivation; population, 40,002; county seat, Opelousas. This is one of the largest parishes in Louisiana. The cattle are in the best condition, and near Estherwood you will find superior grass, comfortable farms on the prairies, and corn, cotton, and even sugar-cane in successful cultivation. Farther west we come to boundless prairies with large herds of cattle grazing on them. Mermentau River Station, 179 miles from New Orleans, can boast of a navigable stream; along its shores are fine farms and beautiful trees. Small sailing vessels and boats are on the river, and a ferry-boat conveys to the other side, which makes here a sharp bend. Cattle-pens tell of cattle-raising and stock farms, and the live-oaks grow to an enormous height. Welsh, also a shipping station for cattle, has cattle-pens, some few stores, a fine station house; many horses and colts, cattle, hogs, etc., roam over the prairies, which are bordered with dense forests. Parish Calcasieu, with Lake Charles as county seat, is 217 miles from New Orleans, with an area of 3400 square miles, but only 14,003 acres under cultivation, and a population of 12,483 inhabitants. The lake is strangely beautiful, and on moonlight nights, with the silvery light scarce penetrating the branches of the gigantic moss-draped cypresses—standing "like Druids of old, gray, indistinct in the twilight"—the scene is fantastic and impressive beyond description. The lake is the resort of the blue-heron and the crane, and the home of the alligator. Your boat crushes pale water-lilies and grotesquely lovely aquatic plants; your oars startle fish that glide in the shallows along the shore. Schooners and other vessels spread their sails on these placid waters; small fishing craft dart on the surface of the lake, scarce disturbing the slumbers of some monster, black and scaly, sunning himself in a spot fit for a fairy revel. At night, when naught is heard but the sough of the wind in the gloomy cypresses, or the faint wash of ripples on the shore, camp-fires gleam like jack-o'-lanterns here and there in the darkness, and the hunter stretches himself in the dancing light and dreams of the weird beauty of the moonlit lake.

TEXAS.

THE outlet of Lake Charles connects with the gulf, and the lake has the largest lumber trade in the state. The lumber, often seen floating in rafts of logs, is shipped to the gulf. Swamps extend to the gulf and partly to the Sabine River, which latter stream we cross in entering Texas. The Sabine River forms part of the eastern boundary, and the Neches River constitutes the western and southwestern boundary. Orange County has an area of 500 square miles, and is about equally divided into prairie and timber land. The soil is very rich; the principal crops are cotton, sugar-cane and vegetables. The soil yields two crops a year, and peaches, oranges, figs and plums have a luxuriant growth. Yellow pine and cypress seem to be inexhaustible along the river, and are of the best quality. Of Orange City very little is known, but it is an enterprising place, and has a great future. In regard to its shingle factories, its lumber trade, and its saw-mills lighted by electric lights, it certainly ranks first in the state. The Sabine River is navigable, and Sabine Pass is destined to become a great harbor. The city is beautifully planned, and is second to no other in Texas in regard to its gardens, orange groves and tropical fruit growth. It is increasing rapidly both in wealth and population (now about 3000). Sabine Lake, ten miles below the city, is about twenty-five miles long and ten miles wide; is navigable and connects with the gulf.

The next county that we traverse is Jefferson, south of Hardin County, with an area of about 900 square miles, and a population of about 3000 inhabitants. The larger portion is prairie, and the rest of the land is covered with a great variety of timber, as oak, hickory, pine, cypress, magnolia, live-oak, etc., finely watered by the Neches River and some of its tributaries, with the well known Taylor and Pine Island bayous. The county seat is Beaumont, on the Neches River, the present population of which is 1500 inhabitants. The town will certainly increase, as the railroad is to be built to Sabine Pass; it is even now in a very flourishing condition. The staple productions of the county are cotton, corn, rice, vegetables of all kinds. Stock raising is an industry that is on the increase. The prairie land is rich, and the climate, although warm, is agreeable on account of the breezes that blow from the gulf.

We now enter Hardin County, about three miles from Sour Lake Station, and a distance of about nine miles from the lakes themselves, if you travel by stage. Hardin is situated in the southeastern portion of Texas; has an area of about 900 square miles, and about 2000 inhabitants. It is comprised within the heavy timber belt of eastern Texas, and only one-tenth of its area is prairie land. The largest portion of the timber is a very fine quality of yellow pine, but you will also find a great variety of oak, hickory, beech, walnut, holly, etc.; the timber is valuable because it can be floated to saw-mills in Beaumont. The chief products of the county are corn, sugar-cane, cotton, peas, sweet and Irish potatoes; vegetables could be raised in large quantities, and fruits of every kind. The climate is fine on account of the gulf breezes. There is an abundant supply of water from clear running streams and creeks tributary to the Neches River, which are of sufficient volume to float the heaviest timber. But the most valuable and most interesting feature of Hardin County is Sour Lake. The surroundings of this wonderful lake have never been improved until the present enterprising proprietor, Mr. Willis, took charge of the place. It is his intention to publish in a very short time an illustrated pamphlet that will treat exclusively of Sour Lake, and for this purpose the ready pen of the well-known editor, Mr. Jack Redmond, has been secured, and the brochure will be highly interesting. The area of the lake is somewhat over two acres, the water is sulphurous, aluminous and ferruginous, and is used for bathing purposes as well as drinking. It is regarded as a specific for rheumatism and cutaneous diseases. For scrofula, skin diseases of any description, it is a complete cure, and will prove to be more beneficial than even the celebrated Hot Springs, Las Vegas, or springs of equal renown. There is an oil that rises upon the surface of the waters at Sour Lake that is possessed of wonderful curative properties, and it has been demonstrated in many cases that it clears and beautifies the complexion as nothing has ever done before. It leaves upon the face the unmistakable glow of health, which can never be successfully imitated by cosmetics. So far, it has been discovered by careful analyses that these waters show about twenty-seven different ingredients; and the experiments during a period of forty years are sufficient to prove that, with a few exceptions, among which is consumption, the Sour Lake water will cure every disease that flesh is heir to. In the immediate vicinity of the lake are numerous wells (at present about fifteen) especially used for either drinking or bathing purposes, according to the directions of attending physicians who have been at the lake for years. We give the analysis of one gallon of the water, made by E. L. Wayne, of Cincinnati, July 14, 1877 :

Free sulphuric acid,	47.25	grs.
Sulphate of iron,	6.92	"
Sulphate of lime,	9.90	"
Magnesia,	4.20	"
Alumina,	7.35	"
Organic matter,	1.23	"

As it has been stated before, there has been made no complete chemical analysis of all the thirteen or fifteen distinct and separate combinations now used, and the distinguished and learned Professor Streeruwitz has been engaged to make an analysis of all of them during the coming season. After an agreeable drive in a comfortable conveyance over prairie and through a lovely forest we pass through Liberty County, one of the southeastern counties of Texas, with about 1100 square miles and containing about 5000 inhabitants. The greater part of this region is level country ; over three-fifths is prairie land and the remainder is well timbered, the trees being of the same kinds as those of Hardin County. The soil of Liberty County is very fertile, and is especially adapted to the cultivation of sugar-cane, cotton, corn, rice, potatoes, etc. The county is about 200 feet above the gulf, and the light breezes from the ocean make the climate very delightful. The town of Liberty is a thriving one, with about 600 inhabitants, and is situated on the beautiful and navigable Trinity. The railroad traverses the entire county. After passing Crosby we cross the San Jacinto River, in Harris County, and reach Houston, the great railroad centre of Texas.

HOUSTON.

From various statistical and other sources of information we quote some facts most interesting to the tourist, who will no doubt spend a day or two in a city whose citizens always distinguish themselves by a cordial hospitality not often equaled elsewhere.

The prominent commercial advantages of this city rank it among the leading business places in the Lone Star State. Houston, situated in Harris County fifty miles from the coast, is at the head of navigation on the Buffalo Bayou, and is reached by tide water from the gulf.

Formerly a line of steamboats plied regularly between this point and Galveston ; now, however, Clinton, a small place six miles below, is the nearest landing for steamers, it being the landing place for Morgan's gulf steamers. It is believed, with reason, that governmental appropriations will soon make this short channel navigable for the largest vessels ; and this, with the dozen railroads centreing here, will make Houston the commanding centre of trade, transportation and commerce.

It was founded in 1836 by the Allens, but has now a population of about 25,000 souls, with a taxable property of over $7,000,000. It is situated at the western verge of the great timber belt of eastern Texas, which stretches from Arkansas to the gulf coast, and on the eastern limit of the great prairies of Texas, which extend to the Rio Grande and New Mexico. It is on the direct line between New Orleans, the metropolis of the South, and San Francisco, the great business centre of the Pacific slope. New life, new ambition, new strength and energy are coursing the arteries of city government. The finest school-houses and the best system of public schools in the state have already resulted. A handsome court house, electric light system, finely paved streets, and the largest hotel in the state are results soon to be realized.

Among the numerous societies the Lyceum, a literary institution, may be mentioned as the leading one, and may be termed a conservatory of music and literature ; it has done much to improve the taste for the fine arts. During the past season it gave a series of most interesting and instructive entertainments. By the generous patronage of the public the directors and members have been very much encouraged.

There are many other organizations, such as Masons, Odd Fellows, Good Templars, Young Men's Christian Association, Knights of Pythias, Heptasophs, Knights of Honor, Young Ladies' Benevolent Association, Independent Order B'nai Brith, Volksfest Association, Turn Verein, Deutsch Gesellschaft, Orchestral Glee Club, Agricultural, Mechanical and Blood Stock Association, Horticultural and Pomological Association ; Houston Light, the first military organization in the state ; Houston Press Club ; six newspapers ; several public parks, prominently the State Fair Grounds and Driving Park, Emancipation Grounds, Brashear Park, Tivoli Garden and Merkel's Grove. All of which go to prove that Houston is second to no city in enterprise and progress.

Besides the above mentioned there are fifteen or more corporations, viz.: Texas Express Co., Pacific Express Co., Houston Direct Navigation Co., Buffalo Bayou Ship Channel Co., City Transfer Co., Texas Immigration Association, Houston Insurance Co., City Street Railway Co., Houston Gas Light Co., Cotton Exchange and Board of Trade, Young Men's Real Estate and Building Association, Railroad Real Estate and Savings Association, Buffalo Compress Co., Odd Fellows' Building and Exchange Co., Houston Compress Co., Grain Elevating Co., Cotton Mills Co., Houston Flour Mills Co., People's Press Co., Volksfest Association, and a few others.

Hotel accommodations now are equal to any city of the same size, and before winter they will be equal to cities of much larger population. The new Capitol Hotel has all of the latest improvements of to-day, and hasn't a superior, in convenience and beauty, in the South. It is fifty rooms larger than the largest hotel in the state, and probably has more exterior exposure than any hotel of its size.

The health of this city is also most remarkable. Official statistics show that the death rate is less than that of many cities that are considered the healthiest in the land.

In conclusion, one other fact may be mentioned, viz.: that some of the finest stores and some of the finest residences in the state are now in course of construction here.

The world is beginning to learn something of the fair land which the adventurous Frenchmen of the seventeenth century overran, only to have it wrested from them by the cunning and intrigue of the Spaniard, in which the Franciscan friars toiled, proselyting Indians, and building massive garrison missions ; which Aaron Burr "dreamed of as his empire of the southwest," and into which the republican army of the North marched, giving presage of future American domination. The dread pirates of the gulf made the islands of the Texan coast their retreats and strongholds ; Austin and his brave fellow-colonists rescued Texas from the suicidal policy of the Mexican government ; the younger Austin accepted it as his patrimony, and elevated it from the degraded and useless condition in which the provincial governors held it ; it spurned from its side its fellow-slave, Coahuila, and broke its own shackles, throwing them in the Mexican tyrant Guerrero's face ; it nourished a small but noble band of mighty men, who made the names of San Felipe, of Goliad, of the Alamo, of Washington, of San Jacinto, immortal. It crushed the might of Santa Anna, the Napoleon of the West ; it wrested its freedom from the hard hands of an unforgiving foe, and maintained it as an isolated republic, commanding the sympathy and respect of the world ; it placed the names of Houston, of Travis, of Fannin, of Bowie, of Milam, of Crockett upon the roll of American heroes and faithful soldiers, and brought to the United States a marriage gift of two hundred and thirty-seven thousand square miles of fertile land.

This gigantic southwestern commonwealth, which can nourish a population of fifty millions, whose climate is as charming as that of Italy ; whose roses bloom, whose birds sing all winter long ; whose soil can bring forth all the fruits of the earth, and whose noble coastline is broken by rivers which have wandered two thousand miles in and out among the Texan mountains and plains—is a region of strange contrasts in peoples and places. You step from the civilization of the railway junction in Denison to the civilization of Mexico of the seventeenth century in certain sections of San Antonio ; you find black, sticky land in northern Texas, incomparably fertile ; and extensive plains which give the cattle abundant living along the great stretches between the San Antonio and the Rio Grande. You may ride in one day from odorous, moss-grown forests, where everything is of tropic fullness, into a section where the mesquite and chaparral dot the gaunt prairie here and there ;

or from the sea-loving population of Galveston, and from her thirty-mile beach, to peoples who have never seen a mast or a wave, and whose main idea of water is that it is something difficult to find and agreeable to taste if one is exceedingly thirsty.

The state has been much and unduly maligned in many respects ; has been made a by-word and reproach, whereas it should be a glory and a boast. It has been guilty of the imperfections of a frontier community, but has rapidly thrown the majority of them aside, even while the outer world supposed it growing more and more away from what it should be. Like some strange, unknown fruit, it has ripened in the obscurity of its rind, until, bursting its covering, it stands disclosed as something of passing sweetness, whereas all men had willingly believed it bitter and nauseous. Texas has suffered much odious criticism at the hands of people who knew very little of its actual condition ; border tales have been magnified into generalities ; the people of the North and

of Europe have been told that the native Texan was a walking armament, and that his only argument was a pistol-shot or the thrust of a bowie-knife. The Texan has been paraded on the English and French stages as a maudlin ruffian who only became sober in savagery, and the vulgar gossipings of insincere scribes have been allowed to prejudice hundreds of thousands of people. Now that the state is bound closer than ever before to the United States by iron bands ; now that, under good management and with excellent enterprise, it is assuming its proper place, the truth should be told. Of course, it would be necessary to say some disagreeable things ; it would even be just to make severe strictures upon certain people and classes of people ; but it would not be necessary to condemn the state wholesale, and to write of it in a hostile spirit. The first impression to be corrected—a very foolish and inexcusably narrow one, which has, nevertheless, taken strong hold upon the popular mind —is, that travel in Texas, for various indefinite reasons, is everywhere unsafe. Nothing could be more erroneous. There is only one section where the least danger may be apprehended, and that is vaguely known as the "Indian country." Hostile Comanches, Lipans, or predatory Kickapoos might rob you of your cherished scalp if you ventured into their clutches; but in less than three years they will have vanished before the locomotive, or, possibly before the legions of Uncle Sam, who is said to be possessed of a strange mania for removing his frontier quite back to the mountains of Mexico. Indeed, this apprehension with regard to safety for life and property in Texas is all the more inexplicable from the very fact that the great mass of the citizens of the state were interested to maintain law and order, and fought the outlaws who found their way among them with bitter persistence. It is true that during, and for two years after the war, things were in a lamentable condition. Outlaws and murderers infested the high-roads, robbed remote hamlets, and enacted jail deliveries; there were a thousand murders per year within the state limits; but at the end of the two years the reconstruction government had got well at work, and annihilated the murderers and robbers. It was a noteworthy fact, too, that the people then murdered were mainly the fellows of the very ruffians who murdered them; shot down in drunken broils or stabbed in consequence of some thievish quarrels. Of course, innocent people were plundered and killed; but then, as now, most of the men who "died with their boots on" were professional scoundrels of whom the world was well rid.

The correct verdict, however, with regard to the present condition of Texas may be summed up as follows:

"A commonwealth of unlimited resources, with an unrivaled climate, inhabited by a brave, impulsive, usually courteous people, who are anxious for the advent of others to share the state's advantages with them; who are by no means especially bitter on account of the results of the war; who comprise all grades of society, from the polished and accomplished scholar, ambassador, and man of large means, to the rough, unkempt, semi-barbaric tiller of the soil or herder of cattle, who is content with bitter coffee and coarse pork for his sustenance, and with a low cabin, surrounded by a scraggy rail-fence, for his home.

"The more ambitious and cultured of the native Texans have cordially joined with the newly-come Northerners and Europeans in making improvements, in toning up society in some places, toning it down in others; in endeavoring to compass wise legislation with regard to the distribution of lands, and the complete control of even the remote sections of the state by the usual machinery of courts and officials; and the binding together and consolidation of the interests of the various sections by the rapid increase of railway lines."

The above is the picture presented to the able writer of that day as he viewed Texas then, while the contents of this brief volume we have endeavored to make a true picture of that section to-day penetrated by the Sunset Route and its numerous branches.

HARRIS COUNTY.

One of the most important counties in the Lone Star State, with an area of 1832 square miles, and with a population of nearly 30,000; situated upon the thirtieth parallel of latitude and between the eighteenth and nineteenth degrees of longitude west from Washington. With about three-fourths prairie-land, the remainder of the county is covered with a heavy growth of timber along the margins of streams, mostly confined to the eastern portion of the county. As ten lines of different railroads cross this county, and it has besides a communication by water from Houston to Galveston, it offers to the public such facilities of transportation as are rarely found. The prairies of Harris County, beautifully undulated, are for the most part devoted to stock-raising; besides, the soils are of considerable variety, (the black, waxy and sandy loam predominating), and can be tilled for generations without deterioration, being well adapted to the growing of cotton, cane, oats, vegetables in great variety and fruit in abundance. The value of all the principal productions increases with the extraordinary convenience to a good market, chiefly to Houston, the county seat, with a population of about 25,000, and at which no less than ten railroads centre. Houston, besides being the centre of all railroads in Texas, is beautifully encircled with groves of live-oaks and stately pines; the whole city is in fact a garden, and roses and other sweet daughters of Flora never cease blooming summer and winter. The Buffalo Bayou, fringed with magnolias, cypresses of gigantic size festooned with mustang grape and other vines, with trees and shrubs of every description bordering its shores, affords facilities for boating rarely found elsewhere. The drives toward Harrisburg, early in the spring, along the bayou, through never-ending natural parks, containing trees covered with vines and draped with Spanish moss, is truly a delightful pleasure. The whole atmosphere is perfumed by the magnolia blossoms; Spanish-daggers and palmettoes deeply shadow the right and left of the road; all around are the woods, a wonderful mass of verdure; and the beholder, enraptured by the sweet song of the mockingbird, will find no more beautiful landscape in America. Besides Houston, there are several progressive towns in the county, among which we mention Peirce Junction, ten miles from Houston, which is the original eastern terminus of the Galveston, Harrisburg and San Antonio Railway, and is an important live-stock shipping point. Spring Station is another thriving place, convenient for lumber shipment; so also Westfield, a small town north of Houston, and several others. The county is well watered by a number of streams, among them

COTTON AND SUGAR-CANE FIELD, NEAR HOUSTON.

the San Jacinto River, Cypress Creek, Spring Creek and others.

FORT BEND COUNTY.

The noble Brazos River winds its yellow waters between beautifully timbered shores, almost doubling its length by its crooked way in this county. It is very interesting to quote Mr. W. P. Quiggs' remarks in regard to the wonderful soil of the heavily timbered bottom land, about six miles in breadth. He has been for years one of the largest sugar planters in this section. He says, " I consider the Brazos and Oyster Creek lands to be the best in the state. They produce abundantly and are very easy to cultivate.

" The land is a rich, reddish, alluvial soil, mixed with small shell. The soil is so deep that I have seen wells dug thirty feet, that at the bottom being as that on the top. The chief products are corn, cotton, sugar-cane, sweet and Irish potatoes.

" The average yield of corn is per acre from fifty to seventy bushels; of cotton, from one to two bales; of sugar, fifteen hundred pounds and ninety gallons of molasses; of sweet potatoes, from two hundred to three hundred bushels; peaches are a safe and abundant crop; plums and grapes flourish luxuriantly. There is an abundance of timber for all purposes, ash, oak, elm, box-elder, hackberry and wild peach. The depth of timber from the river to the prairie will vary from three to five miles. The prairie furnishes plenty of fine grasses for summer pasturage, while the timber bottom lands afford protection from the ' Northers ' and bad weather, with plenty of grass all the winter. I have been planting for six years, and I make this statement from actual experience. I have never found any difficulty in procuring all the labor I wanted, and have taken off four hundred bales of cotton, two hundred hogsheads of sugar, and corn without limit. Corn has ready sale at fifty cents."

The beautiful prairies, interspersed with lovely groves, dotted with herds of cattle, almost hidden in the ever-living grass, have to be *seen* and *traversed* that one may comprehend the beauty, and breathe the exhilarating air of an ocean of flowers.

Oyster Creek, a slowly running stream of a dark color, influenced by the tides of the gulf only during dry weather, runs nearly parallel with the Brazos. Very extensive cane-brakes, existing formerly on its banks, are now partly destroyed by cattle and by cultivation of the soil by industrious colonists, now rich and independent citizens. Richmond, thirty-four miles from Houston, is the county seat of Fort Bend, and is situated on the west bank of the Brazos. It has two railways, the Galveston, Harrisburg and San Antonio, and the Gulf, Colorado and Santa Fe. The county is north of Brazoria and south of Harris counties, in longitude eighteen and nineteen degrees west, and in latitude twenty-four and thirty degrees north. Area, 889 square miles, and population about

10,000. The health of this county, so near the gulf coast, is excellent, the inhabitants are refined, intelligent and industrious. In regard to the products that were exhibited at the pomological fair held at Houston in 1878, they were unexcelled, comprising sugar-cane, corn, cotton, fruits and melons of every kind, immense in size and excellent in quality. The most wonderful articles on exhibition were thirty-six different varieties of *native* nutritious grasses; fifty-four varieties of timber cut from the forests of the county; the finest and best honey, and specimens of rock and different soils. The railroad next traverses a portion of the northern section of

WHARTON COUNTY.

South of Austin County and north of Matagorda, this county lies between the twenty-ninth and thirtieth degrees of north latitude, and on the nineteenth degree of west longitude, with an area of 900 square miles, and a population of about 5000, mostly white. It is watered by the Colorado flowing through its centre, with Old Caney and East and West Bernard Rivers. Peach, Middle Bernard, Lone Star, East, West and Middle Mustang, Pin Oak, Golden Rod, Sandy, Jones, Blue and Palocois Creeks, traversing every section of the county, makes it the best watered along the Sunset Route, and therefore not surpassed in productions by any other portion of our state. The soil is the rich alluvial, and the county is divided by lovely valleys, well timbered, and prairie lands awaiting the cultivator; for as yet only a small part of the county is under cultivation. The timbered portion abounds, like Fort Bend County, in every variety of oak, cypress, cottonwood, ash, pecan, elm, etc. The crops are principally cotton, corn, sugar-cane; and it is stated that the corn is better than corn raised in the East or North. Cattle and other stock are in excellent condition both winter and summer, as they find grass, other provender and water during all seasons. Horses are herded, and it is estimated that from five to six thousand are in the best condition, grazing on the prairies with many thousands of Texas cattle. The principal towns of Wharton County are Wharton, the county seat, situated on the Colorado River, with a population of about 600, and New Philadelphia, a splendid location recently laid out by Pennsylvanians, sixty-three miles from Houston. This place is growing rapidly, and yet there is much room for immigrants who desire a home for themselves and children. Eagle Lake is a favorite camping place. It is one of those lovely spots which never fail to attract the tourist and those who seek seclusion from the fatigue of city life. The hunter, early in the morning, is sure to find ducks, geese of various kinds, the plover, the snipe and curlew, or the sand-hill crane and the trumpeter crane; the latter is a beautiful bird, standing quite six feet high, with white plumage relieved by black wings and back. The angler is sure to get a bite of a black bass (trout), white perch, catfish, or a buffalo ; the latter, similar to the European carp, has a very fine flavor. The botanist or florist will find water roses and water lilies with other beautiful specimens of aquatic plants peculiar to this lake only. Unimproved lands, mostly owned by some speculators, are sold from one to ten dollars per acre, while cultivated farms range from fifteen to forty dollars per acre. Some new settlers, about four to five years ago, have, in this locality and about Schulenburg, founded colonies which are improving wonderfully ; most of them are composed of Germans. May they prosper!

COLORADO COUNTY.

This county, traversed by the Colorado River and the Galveston, Harrisburg and San Antonio Railway, has a splendid reputation in regard to its free school system, as the citizens take a warm interest in educational matters. They have built fine school-houses, etc., and although anxious and generous to promote general welfare, they are free from debt—a sure index to prosperity. On the west bank of the Colorado River is situated Columbus, the county seat, eighty-six miles from Houston. This city is one of the most charming along the line. Along all the streets fine rows of shade trees of different varieties are planted, and the taste formerly displayed by the first settlers in preserving the exceptionally fine live-oaks has given this lovely place a poetically romantic appearance, and they may justly take pride in having the finest specimens of live-oaks in the state. Strangers passing through the city admire these grand old trees. Draped with the gray Spanish moss, during moonlight evenings they are enchantingly beautiful in effect and form—the more so when the gentle gulf breeze, with which they are always favored, gracefully swings the long, waving, ghost-like hair of these kings of the former forests on the Colorado River.

In regard to health, Columbus is a favored spot on earth, as the nights during summer are made cool and pleasant by a regular breeze from the gulf. Persons whose constitutions are worn out are there restored to new life and happiness ; even cases of consumption have been cured, and patients have regained perfect health. As the country throughout the county is undulating, and as the rivers and creeks are running toward the gulf, no pools of stagnant water can collect to produce miasma.

Although there is a predominance of prairie land, there is enough timber for fuel, building and fencing purposes, such as post-oak, water-oak, burr-oak, live-oak, black-jack, hickory, pecan, cypress, elm, ash, walnut, cottonwood, willow, sycamore, etc. The bottom lands are excellently adapted for the cultivation of corn, cotton, all kinds of grain, fruits and vegetables. Besides the Colorado River, the county is watered by the Navidad River, Harvey's, Cummins, Skull and Sandy Creeks.

FAYETTE COUNTY.

It is located in longitude 20 degrees west and latitude 30 degrees north. The Sunset Route passes through this county, also the Colorado River, with numerous creeks traversing the same, which water-courses afford an inexhaustible supply of water for all purposes. The names of these creeks are : Peach, Live Oak, Pin Oak, Buckner, Barton, Cedar, East and West Navidad, Mulberry, Rocky, Middle, Williams, Criswell, Babbs, Jones, Cedar No. 2, Clear, High Hill, Owl, Cummins and Haw Creeks. This county was organized 1838, and has an area of 975 square miles. About one-half of this county is tillable prairie, one-fourth tillable timber land, and one-fourth first-class timber land. The surface of the county is rolling, and, in regard to landscape scenery, a

paradise for artists who are in search of the picturesque. The whole country is alive with stock, cattle, sheep, hogs, goats, mules, horses —found along the creeks or traversing the prairies that are ornamented with beautiful shrubs and groups of trees. The two lakes, Primms and Crownover, are the homes of all kinds of water-fowls, fishes, turtles, etc.

The timber is principally post-oak, but there is along the creeks and lakes a limited variety of other kinds. The soil is waxy, black and sandy loam ; it is very productive and bears the highest cultivation. The excellent facilities for transportation, still increased by another proposed railway, adds considerably to the value of the land, and, considering all advantages, renders this county one of the most desirable in our state. It is attracting a large immigration and has some of the richest and finest settlements of this state—due to the energetic, intelligent and industrious inhabitants. The climate is mild and healthy ; the principal agricultural productions are cotton and corn. The stock-raising interest of this county is of importance, and statistics prove that over 70,000 of various kinds of domestic animals are registered. The county has an intelligent population of about 30,000. There are ninety-eight schools, fifty churches, six Masonic lodges, and various other societies. The Galveston, Harrisburg and San

Antonio Railway passes through some of the most important cities of this county, viz.: Borden, Weimar, Schulenburg and Flatonia.

La Grange, the county seat, with a population of over 3000, is connected by a branch railroad with Columbus and the main line. Next town in size and of importance is Schulenburg, 111 miles from Houston, with a population of over 1200; then follow villages, each one the home of happy citizens of different nationalities, such as Fayetteville, Ehlinger, Round Top, Warrenton, Ledbetter, Rusterville, Winchester, Cistern, New Prague and High Hill. The water power of the Colorado River will soon become utilized; it needs only capital and energy to convert the bending Colorado into a powerful motor that will move a hundred industries.

GONZALES COUNTY.

This county, partly traversed by the Sunset Route, ranks among the very best agricultural counties in the state. The county is about sixty miles in length and about twenty-five in width, containing 1100 square miles; with a population of 18,000, rapidly increasing. In regard to scenery, water-courses, springs, soils, etc., this county has a great variety, and we only attempt to give some general outlines of such a vast body of land.

The Sunset Route, in traversing this county, passes through Waelder, Harwood and Luling—the latter place one of great importance as a winter resort for invalids, of which we will speak in its proper place. Gonzales is the county seat, and situated on the Guadalupe River, about a mile below the mouth of the San Marcos. It has over 2000 inhabitants, very fine buildings, a stone court-house, a large college building, etc., and offers to immigrants great inducements.

There are few counties that are better watered. The Guadalupe and San Marcos Rivers, Peach, Plum and Sandy Creeks, and many rivulets, flow through it; besides, there are never-failing lakes well supplied with good water, even during the driest seasons, and in all portions of this county well-water is found by digging from twenty to sixty feet. Most of the springs, and both rivers, contain lime-water—cool, healthy and very palatable after a short use of it. Some of the wells and springs are sulphur and some sour; but near-by plenty of lime-water (sometimes freestone) can be found. The land upon these rivers and creeks is naturally a fine alluvial soil, covered with a fine growth of black-walnut, burr and Spanish oak, hackberry, mulberry, pecan, cottonwood, elm, ash, willow, sycamore, alder, box-alder, etc., and a dense undergrowth of black and red haw, buckeye, wild china and plum, dogwood and dogberry, and many small vines. The mustang grape-vines twine almost every tree and swing in graceful festoons from their boughs, heavily laden with rich clusters of grapes. Prairie land forms the higher ground, in which nature stored away inexhaustible

fertility, with a fine growth of native grasses, affording constant pasturage in spring, summer and winter. The uplands, touching these prairies, are well timbered with post-oak, live-oak, black-jack, hickory and various grape-vines; and are also interspersed with undulating prairies with rich black soil covered with mesquite grass and mesquite timber.

This is certainly the spot where the poor man may live independently, or the place for the capitalist to increase his wealth; the more so as few healthier spots can be found in Texas. The climate is mild, the thermometer rarely getting higher than 96 degrees or lower than 25 degrees, and during summer a healthy gulf breeze renders the country pleasant everywhere. Corn, cotton, oats, rye, wheat, millet, tobacco, sorghum, ribbon-cane, melons, peas, beans and all kinds of garden vegetables can be raised in abundance. Peaches, all kinds of grapes, plums, pears, figs, apples and apricots grow there; and as there is an abundant supply of post-oak, pin-oak, pecan and black-jack, the mast alone is sufficient to fatten hogs.

The new settler may rent land and will be furnished everything necessary to make a crop, and get one-half what he makes; or he can buy land (the best in the county) from $1.50 to $5.00 per acre.

The citizens of this county are enterprising, liberal, intelligent and industrious, inviting the immigrant and others to share their resources. From Gonzales the railroad passes into

CALDWELL COUNTY,

"with the charming valley of the San Marcos River, renowned in song and story;" "its rich and fertile soil, its stately cedars and towering pines, its bloom and beauty and fragrant summer breezes—an Eden on earth." Caldwell County has an area of 522 square miles, about two-thirds of which is timbered and one-third prairie land. The timber consists of post-oak, elm, walnut, ash, hickory, mesquite, etc. The prairie lands are, as usual, very rich, and produce abundantly cotton, corn, vegetables, many of the cereals, fruits, grapes, etc., as do the lands in Gonzales County. Lockhart, with a population of over 1000, is the county seat. The railroad passes through Luling only; some miles from the road is the charming village, Prairie Lea. The total population of this county is over 10,000, and the gratifying increase is owing to a large immigration here by the Galveston, Harrisburg and San Antonio Railway, and consequently much new land is becoming cultivated. The county swarms at present with all kinds of stock, and new homes give shelter and plenty to all settlers.

The springs of this county are celebrated for their medicinal qualities, and are visited by thousands who are seeking health—Burditt's Sour Wells, Cardwell's Spring, and others not far from the Sunset Route. As a winter resort these springs present unusual attractions, and together with the charming scenery and delightful climate they excel by far the favorite watering places upon the southern Atlantic coast. Therefore, some general outlines of the most prominent mineral spring near the Sunset Route will be given herein.

LULING SPRINGS.

"These springs have come to be of marvelous value during the past few years, and they have the advantages of good hotels and places where invalids can be properly and comfortably cared for. Luling is situated in Caldwell County, on the line of the Galveston, Harrisburg and San Antonio Railway, 155 miles west from Houston and 57 miles east from San Antonio. The reputation of the Luling sour water is wide-spread, and large numbers from the North and East are seeking its benefits. It is also shipped to different parts of the country, and furnished to those who cannot afford the expense of visiting the springs. As a winter resort Luling is one of the most desirable in Texas. The climate is like that of San Antonio, and in addition to the great value of its waters is the mild and invigorating atmosphere, which alone will restore health to a system that has become weakened and depleted. The most alarming cases of inflammatory rheumatism, and indeed all diseases of an inflammatory order will yield to the use of these waters. It is also a specific for the cure of chronic diarrhea, and many other of the diseases that afflict mankind.

"The following is the analysis of the Luling sour water : Sulphate of lime ; sulphate of magnesia, in large quantities ; chloride of sodium ; chloride of potassium ; carbonate of iron, in moderate quantity ; phosphate and choride of lime ; sulphates of alumina and baryta, and traces of silica and strontia, and large quantities of free sulphuric acid. Its benefits extend to all diseases that proceed from an abnormal condition of the biliary secretions."

In Caldwell County have been discovered traces of silver ore, and southeast of Lockhart an abundance of iron, while in other parts veins of coal have been found ; all of which will be fully examined and utilized as soon as this beautiful county becomes more thickly settled. Now to the celebrated

GUADALUPE COUNTY.

The beautiful San Marcos River forms its northwestern boundary, the clear Cibolo runs through its western section, while the noble and charming Guadalupe River traverses its central portion, forming one of the loveliest valleys in Texas, and not only the loveliest, but the most productive. The general character of the surface of this county is undulating and varied. It is situated on the twenty-first degree of longitude west from Washington, and between the twenty-ninth and thirtieth degrees of latitude north. The soil ranges from the rich, black, sandy loam of the mesquite lands, to a light, sandy soil which is easily tilled and of great productive strength ;

especially so along the streams, where a large percentage of humus adds greatly to the fertility.

The rapids of the Guadalupe River, not far from the celebrated iron bridge, are beautiful ; the clear water dashes over a rocky plateau and forms many graceful cascades ; between these cascades numerous small islets, covered with shrubs, relieve the eye from the brilliant white foam of the Guadalupe ; anglers are engaged in securing trout and other specimens of the finny tribes with which this river abounds. Looking up these water-falls the iron bridge is suspended over the river ; on the right shore of the Guadalupe, majestic cypress trees stretch their graceful branches over numberless other trees and shrubs of smaller size. The whole landscape is all light and shadow—the dark green shores contrasting finely with the pearly hues of the foaming cascades ; and over-arching all is a vault of deep azure, where the white crane sails and the sinking sun reflects in all its splendor the last rays in the sparkling river.

Wood-land and prairie about equally divide the county, which is comprised within an area

of about 800 square miles, with a population of over 11,000 intelligent and thriving citizens. Its elevation above the gulf is 700 feet, the mean temperature about sixty-nine degrees, the rainfall sometimes thirty-four inches. The products are cotton, corn, oats, rye, wheat, all kinds of vegetables and fruits, also grapes in abundance. Timber of but moderate growth is found here, but enough for domestic or fencing purposes. Along the rivers are fine sections of oak and black-walnut timber ; and as springs and creeks abound in every section in this county, everything grows luxuriantly. Seguin, the county seat, is a flourishing town of 2500 inhabitants, situated upon the north bank of the Guadalupe River, on the line of the "Sunset Route," 35 miles east of San Antonio, and 180 from Houston. The town is built upon a beautiful plateau, interspersed with giant live-oaks, beneath which are many sparkling springs. A tributary of the Guadalupe passes through the city, which is spanned by two fine bridges. The public buildings are large and comfortable, and besides many elegant private residences there are ten churches, the college and a high school. In and near Seguin are three water-mills, and west, about one mile below the railroad bridge, is a fall of ten to twelve feet, where several hundred horse-power might be yet utilized.

The climate so near San Antonio is delightful ; stock-raisers do well, and to the agriculturist rare inducements are offered. Next the Sunset Route enters the world-renowned and historical grounds of Bexar County, but passes first a village by the name of Marion, 191 miles from Houston, which place is well known, as the stage-coach takes from there passengers to various inland villages.

BEXAR COUNTY

has an area of 1475 square miles and a population of over 32,000. The county is situated in the southwestern portion of the state, between the 29th and 30th degrees of latitude north, and between the 21st and 22d degrees of longitude west from Washington. Formerly this county embraced an area larger than the State of New York, and is perhaps the best known county in Texas. Its topography is a grand undulating prairie, a portion of which is timbered with varieties of trees usually found in this section of the state. The altitude is high ; the soils are very rich, ranging in depth from three to fifteen and twenty feet. The county is watered by the Cibolo, San Pedro, San Antonio and Medina Rivers; the Leon, Medio, Colabras, Cottonwood, Balcones and Geronimo Creeks, and a large number of springs which flow from the base of limestone and sandstone formations, including the famous San Pedro and San Antonio springs.

This county, under favorable circumstances, will produce abundantly cotton, corn, wheat and the other small grains ; also tobacco, rice, sugar-cane, broom corn, castor beans, California clover, Hungarian grass, millet and every variety of vegetables ; and of fruits, such as peaches, pears, apples, plums—while figs and grapes are produced to perfection. The altitude of Bexar County above the gulf, its supply of clear and running water, the cooling breezes from the south, the delightful and exhilarating climate, and its freedom from low bottom lands, renders it the most healthy in the state, and perhaps in the whole country.

In regard to San Antonio, the county seat of this county, with a population of about 25,000, so much has been written as to how it can be reached from Houston or Austin and other parts of the country, and what wonderful attractions the city offers to tourists and invalids, etc., that we prefer the following beautiful description by Mrs. Harriet Prescott Spofford to our own ; we, in addition, will only continue the description where this charming writer stopped some years ago. The following article (ending on page 40) is herewith copied, by the courteous permission of Harper Bros., from their popular magazine :

" At the moment that you start westward on the Sunset Route, the landscape salutes you in all the loveliness of a blossoming prairie in its first luxuriance of green under the tender early sun. The flowers are numberless. When you have counted a couple of dozen varieties, you find you have only begun. Here the painted-cup makes the great reaches gay ; here yellow indigo stars them, and presently lends them its color, leading away into the boundless horizon a Field of the Cloth of Gold ; here it is scarlet with the scarlet phlox, here blue with the verbena ; here the lilies, with their long snowy filaments wondrously alive, whiten all the windings of an unseen brook ; here, clothed in the priceless small clover, and greener than Dante's freshly broken emeralds, beneath vast and hollow heavens, and ' moulded in colossal calm,' the naked prairie rolls away, league after league, unbroken to the gulf.

Oh, the glory of a Texas prairie under a vertical sun ! the light, the color, the distance, the vast solitude and silence, the limitless level, the everlasting rest ! A flock of white cranes rise flashing in the light and soar away ; a mirage lifts the lofty timber that outlines a distant river, and shows you the stream shining beneath, shaking silver vapor at its feet ; in the creek beside you, fearless blue ducks dip and dive and skim away, scattering the water-drops ; a drove of horses, rising from beds of sunflowers, with flying manes and tails, go bounding into space ; vast herds of cattle crop the clover without lifting their heads as you sweep by ; riders are rounding up their droves, hawks are hovering, birds are singing, winds are blowing, and what seemed only solitude and silence is full of life and action and music. Now the forests of the Brazos begin to rustle ; cypress and magnolia, linden and locust, ash and beech and elm, hickory and black-jack, dense to darkness, yet trembling with dew and sun, laced with gay vines of trumpet and passion flowers, and with huge ropes of blossoming grape slung from tree to tree, thick with undergrowth of dogwood and redbud, wild peach and cane, and their great dark live-oaks wrapped in the fantastic shadows of a thousand gray swaying cobwebs, and standing weird and awful in their Druidical beards. And out on what bottom-lands you come—the Nile-rich bottoms of the Brazos and of the Colorado ; the black mould and the chocolate of an unmeasured depth ; the cotton springing in endless rows of opening, bean-like leaves ; the delicate sugar-cane just shaking out its ribbons ! Here in the Brazos we dash by a sugar plantation, the low house with its broad verandas and wide-open doors under huge trees, in the distance the great sugar mills, and all around it the two thousand acres that make it a yearly return of one hundred thousand dollars. In the old times it was worked by a couple of hundred slaves ; now seventy convicts under an armed and mounted guard do as well.

There is, however, let us say in passing, no trouble about work in Texas. Political difficulties were over there sooner than elsewhere

SAN PEDRO PARK.

in the South, and the affairs of labor equalized themselves to the laws of supply and demand. Throughout the state the freedmen are industrious and quiet, securing a good livelihood and laying up money.

Not far away, and still in the Brazos bottoms, by Oyster Lake, a Massachusetts colony is setting up its tents ; and here land may be had for five dollars an acre, better than lands on the Illinois alluvial for fifty dollars an acre, and quite as healthy.

Still we roll on, slowly mounting the six hundred feet of altitude at which San Antonio lies above the sea, out upon other prairies, where a single pasture of one hundred thousand acres fences in its tremendous herds. Flocks of birds darken the air like clouds of leaves, and vanish ; a deer, perhaps, bounds by ; a great buzzard is spreading his ragged wings over his unseen quarry ; a carriage and pair go gently along the springing sod—strange anomaly, so far it seems just then from usual life. We roll past Bernard, whence, with but one house in sight, nearly ten thousand bales of cotton are dispatched ; past the young city of New Philadelphia ; past the lovely Eagle Lake, with its fish and game ; past Schulenburg, whose former owner, annoyed by the approach of civilization with the railroad, refused to sell a right of way, but disposed of his whole estate, and moved on where no one could elbow him—six years ago a homestead, to-day a town with mills and workshops, and daily paper ; past Luling, with its neighboring Sour Springs, working cures by repute wonderful as the Pool of Bethesda, the gate to the San Marcos, whose fairy-like beauty has been so fitly sung by Mrs. Davis, the sweetest singer of the South ; past a score of neat villages, under their live-oaks and pecans ; and so on and up and out on the great grain region of the world, where the tender wheat is springing in long stretches vanishing from sight, the rye is already high, the corn is up two feet—a vast rolling region of plains and sun-bathed slopes, before which Mesopotamia is a fable, and the wealth of Odessa is but dust. Over the Guadalupe then, the three Santa Claras, the Cibolo, the Salado, straight into the sunset that casts out its long beams and reddens sky and prairie, and wells up in a flood of lustre suddenly extinguished by the quick-descending night. Lights begin to twinkle below, and you descend into San Antonio. There is a crowd of dark faces at the station, a confusion of strange tongues. As the carriage goes along, soft wafts of balmiest fragrance salute you ; you are conscious of being in a world of flowers. As you alight at the Menger, enter a narrow, unevenly stoned passage, and come out upon a broad flagged court-yard, surrounded on three sides by open galleries, with the stars overhead, and the lamp-light flaring on a big mulberry-tree growing in it below, you feel that you are in the heart of Old Spain.

San Antonio is like nothing so much as the little African town of Blidah that Eugene Fromentin comes upon in the midst of the desert, set behind jalousies, among gardens and fountains, smothered in roses, and sung to by nightingales. On a more enchanting spot the eye of poet never rested. There is probably nothing like it in America. Four days ago you left the snow of March or April under the windows at home, now your room is full of roses ; and as you go out and about, you find the town one wilderness of roses, a very Vale of Cashmere. Blush and creamy and blood-red, the delicate little Scotch rose, the superb Marshal Niel, the shining Lamarque, the beautiful great tea-rose, hundred-leaved and full, waxen-white, and damask, the heavy-headed Persian rose itself —they hedge gardens by the quarter of a mile together, lattice every veranda, climb and lie in masses of bud and blossom on every roof. It is a long red roof usually, that, bending slightly, forms also the roof of the veranda. Most of the houses beneath it are long and low and narrow, of a single story, and but one step from the ground, built of a cream-colored stone that works easily and hardens in the air, and so placed that the south wind or the east shall blow in every room—the wind that blows all day long from the gulf, and makes the fervent heat itself a joy. There is no vestibule ; you enter the saloon from the door, and the other rooms open on either side of that, and as they all open on the veranda, that is used as a hall. Over them rise the tall cotton-woods and the huge spreading pecans, and before them or behind them, almost invariably, flows a swift, clear, artificial stream of water some four or five feet wide, the banks now stoned in, now covered with a lush growth of the blooming cannas and caladiums, immense arrow-headed leaves the size of an African warrior's shield, and now bridged beneath honeysuckle arbors.

These charming dwellings stand with little regularity or uniformity, but here and there, facing this way and that, just as the winding roads wind with the winding river, and always half buried in a sweet seclusion of leaf and blossom. Not roses only, but all the other flowers under heaven: lilies and myrtles and geraniums make the air a bliss to breathe ; aloes sit drawing in the sunshine, suddenly to shoot out in one long spike of yellow bloom higher than the house itself ; the Spanish-dagger lifts its thick palm-like trunk, and bristles at a thousand points around its great cone of creamy bells ; the euphorbia clothes its strange and lofty stem with a downy green, and then flowers with a blossom like a redbird just alit ; in every vacant space the acacia 'waves her yellow hair'—the very acacia, it is said, with whose long scarlet silken stamens tumbling out of their yellow hood Moore has taken such poetic license. There are groups of bananas, too, the arch of whose huge leaves reminds you only of Paul and Virginia's home ; there are walls of the scarlet pomegranate, one blaze of glory ; lanes lined with the lovely-leaved fig-tree, where the fig is already large; and the comely mulberry-tree, grown to an enormous size, is dripping with its blackening and delicious fruit. Sometimes there are summer-houses at the gate almost half the size of the dwelling, entirely covered with vines, and the whole spot so sequestered behind mimosa and cacti and huge-leaved plants that it seems only a tropical tangle that you might hesitate to enter; but, pushing your way through which, you will find, behind broad porches, lofty rooms with polished floors and rugs, books and pictures and vases and costly furniture, inhabited by white-clad women whose manners have peculiar grace.

In and out among these houses slips the San Antonio River, clear as crystal, swifter than a mill-race; now narrow and foaming along between steep banks rich with luxuriant semi-tropical growth, and with the tall pecans on either side meeting above them in vaulting shadow; now spreading in sunny shallows between long grassy swards starred with flowers, twisting and turning and doubling on itself, so tortuous that the three miles of the straight line from its head to the market-place it makes only in fourteen miles of caprices and surprises, rapids and eddies and falls and arrowy curves, reach after reach of soft green gloom and flickering sunshine, each more exquisitely beautiful than the other. Around every lane it takes a loop; here it is just a pebbly ford, there, although so perfectly transparent that you can see every flint in its bed, it is of a profound depth, and everywhere it is of a color whose loveliness is past belief. It flows by the Mexican jacal, and through the wealthy garden, around the churches, across the business streets with its delightful glimpses. You cannot escape it; you think you have left it behind you, and there it is before you, hurrying along to the forests on its two hundred miles to the gulf. It is a happy course this river runs to-day, but a hard fate is in store for the lovely San Antonio. All its pretty, boisterous play is presently to become the groaning labor of a slave, for the sixty feet of its fall, if it is something to delight the heart of a poet, is something also to dilate the bank account of the manufacturer.

The San Antonio is joined in the valley by the San Pedro, another limpid stream, that pours from the rock and winds through some public gardens before making itself more useful.

The town lies in its valley in the broad basin of the great hills, and upon both sides of the river, and the serpentine course of the river, crossed by a score of bridges and as many fords, is such a confusion and a snare that you never know upon which side of it you are. The streets in the old part of the city are exceedingly narrow, and by no means clean, and the sidewalks are narrower yet, and worn in ruts by the tread of numerous feet. Many of the buildings on this street are of adobe, all of them a single story in height, most of them

with galleries, as the veranda and piazza and porch are called. Some of them have a curious front, the wall projecting a couple of feet above the line where eaves should be, and pierced by rain-spouts, forming a breastwork behind which the defender lay protected, while through the rain-spouts firing down into the streets, which, in the furious old times that this town has known, with now one master and now another, were wont to run with blood.

Narrow as the streets are they are incumbered in every way and made still narrower. Here the incumbrance is carts full of huge blocks of unhewn stone, which are handled by brawny Mexicans and negroes, without derricks, and which the citizens patiently submit to see cut in the streets day by day instead of in the stone-cutter's yard; here it is trains of

clumsy Mexican wagons covered with canvas and drawn by oxen whose yokes are bound upon their horns, thus occasioning every jolt to jar the brain, and shortening the term of service of the stoutest beast. Often the main plaza is entirely covered with these teams, the great oxen lying all day in the sun there, and from under the canvas of the wagons protrude a crowd of little dark faces that make one fancy all Mexico is on the move. Sometimes the incumbrance is a string of donkeys that trot through the streets, each one with a single fagot on his back, oddly contrasted by another where each one is so hidden by his load of straw, hay, fresh grass, sugar-cane, or corn, according to the season, whose long blades and stems trail upon the ground, that only his head and ears show how the bundle moves. Now it is a Mexican family transferring their altar—the Lares and Penates—on a cart, the father leading it, the mother and grandmother totally obscured by the things they lug along, an infinity of children round their heels, dirty and ragged and with tangled hair, but with the blackest eyes and whitest teeth, the ruddiest dark cheek and most roguish smile ever seen, and with the baby all but bare, strapped on a blanket on a mule's back, sound asleep in the sun, as sweet a little morsel as the first baby ever born in paradise. If it is a Mexican family in a cart encountered thus, the mother is always on the front seat, while the father sits behind and holds the baby. Here it is an army train that stops the way, and makes a prominent feature of the streets—huge covered wagons drawn by mules four abreast, with an armed and mounted escort, whose rifles and broad cartridge belts mean business—on its way to the yet distant frontier, between which and the town a train is almost always moving, as supplies are being dispatched, or officers' families are taking their long ambulance journeys. These streets afford a good deal of interest, and add much variety and vivacity to life for the invalids who visit San Antonio seeking health, the number of whom is large, since the air there and in the surrounding region seems to have peculiar properties that render it almost a specific for consumption and diseases of the throat ; and the invalids who have come down there simply to prolong life have in uncounted cases gone away entirely cured. You will see these re-born people, themselves a sight, strolling and driving about in all the pleased surprise of their return to life, and that in a town of such strange and foreign sights to them. Here comes a gay Mexican rider, too, who, if he is in full dress, wears his dark trousers buttoned up the outside of his leg with silver bells ; his jacket rich with dollars, and his belt ; his great light felt sombrero stiff with embroidery of gold and silver ; and his bridle and saddle, stirrup and spurs, shining and clattering again with silver. Or perhaps it is a party of ladies bounding along, for every woman in San Antonio is a fine and fearless rider ; or some heavy cavalry riders, superb in blue and gold ; or else it is a mounted beggar, who, if he does not have a servant to carry his bag, as the Fayal beggar does, yet rejoices in a stout little *burro* of his own. Here on the sidewalks, beneath an umbrella-tree that sheds abroad powerful fragrance, little tables are spread, where the market people get their roll and chocolate and bit of pastry, sitting where the gutter would run if there was one. Here, too, are the venders of strange dark candies, from which the flies are brushed with a cow's tail ; of porcupine-work ; of bunches of magnolias, and great, ineffably sweet Cape jasmines from the coast ; and Mexican women crouch upon the hot stones, their dark sad faces half veiled by their ragged ribosas, surrounded by wicker cages full of mocking-birds, vivid cardinals with their red crests, and lively little canaries on whose plumage every color under the sun glistens, making the tiny creatures marvels of emerald and gold and ruby and turquois. These Mexican faces are a great part of the little town ; there are portions of it, called Chihuahua and Laredo, where you see nothing else. There, tumbling in the dirt, are the Mexican babies, than whom nothing can be lovelier ; there, too, are the Mexican grandmothers, than whom nothing can be uglier. Here you can buy skins of leopards and ocelots, which the Indian women dress with the brains of the beast till they are as supple as silk ; here are the little Chihuahua dogs that can nestle in the sleeve of your coat ; here is wonderful Mexican needle-work, made on the drawn thread, rivaling the Old-World laces ; here are earthen pipkins or *jarritos* prettily ornamented, with their *molinillos*, or curious wooden sticks, set in many rings, which, rolled upright between the palms, make the chocolate foam in the pipkin. Whatever you buy, *pelon* will be given you ; and whatever the Mexican buys himself, be it but five cents worth, he expects *pelon*, or something thrown into the bargain, which renders him not too profitable a customer. Here, in these old regions of the town, you can still see the women patiently crushing the corn on the matata ; and here, at almost any hut, you can get Mexican refreshment, if you wish it, that will make you odorous for days.

Everywhere about the outskirts of the town are innumerable low huts built of sticks and mud and straw and any old drift, roofed with thatch coming almost to the ground, and presenting an appearance of the utmost squalor. These are the Mexican jacals. The chimney and its ovens are usually in a cone of baked and blackened mud a little removed, and under a rude awning or a tree the whole family is usually to be seen, with mules, donkeys, chickens, and a horde of dogs, among the latter a hideous, hairless animal, promiscuously intermixed. Dogs are largely in the majority of the population in San Antonio, and their baying divides the noises of the night with the cock-crowing that resounds from house to jacal, from farm to ranche, and rises on the ear in broad surges of sound like the waves of the sea. If you should glance into one of these jacals, you would find an earthen floor cleanly swept, a bed neatly made and brightly covered, and the place garnished after its sort ; and although the general idea is that it is a nest of filth, to the casual eye it seems clean and orderly, but poor to the last degree of poverty. Yet the Mexican here can live on less than any. In the summer the corn and onion and peppers of his garden-patch meet his needs ; in the winter, even when he owns his bit of land, a five-penny soup bone and one sweet potato comprise his usual marketing. But poor as he may be, his daughters do not go out to service ; his mother wraps her ribosa—that remnant of the Spanish mantilla—about her with the grand air ; and he himself, although in rags, salutes you on the street with the grave courtesy of a Spanish don. Making exception of the proud old Mexican families of lineage and repute, who live in seclusion, it is not possible to feel that these people who are known as Mexicans have any claim to the name as we use it. They are simply a gentler Indian, accepting a sort of civilization, now and then with a fairer tint, now and then with a wave in the hair that tells of darker blood, and always with a high cheek-bone, following them to the tenth generation. The proud Castilian has but small part in them, the gentle Montezuma race perhaps has

MISSION LA CONCEPCION.

less. One having those two strains in his veins—the Spaniard, with his hemisphere of poetry behind him ; the Montezuman, representing ancient and rightful empire of the continent—should wear, it would seem, other than these low-browed faces stamped in their dumb and sullen ignorance, whether you see them on the women squatting on the brick floor of the cathedral, or on the men lounging in the plazas against anything which will uphold them, darker and more sullen for the shadow of their huge sombreros.

San Antonio is, in fact, a Spanish town to-day, and the only one where any considerable remnant of Spanish life exists in the United States. In its old archives much interesting information is held concerning the early Spanish rule in this country, and here also, by-the-way, are some papers going very far to prove the utter innocence of Aaron Burr of the treason under the charge of which he suffered. Many of the people proudly call themselves Spanish, and most of the Americans of the region find it necessary to speak their tongue easily ; a lawyer, indeed, could hardly practice his profession without knowledge of the language, which he needs in examining witnesses, in pleading, and in recourse to the documents in the matter of land titles, many of which are in the Spanish, while most of the local laws are founded on old Spanish usage. Land is still measured here by the vara, and the town has its alameda, its plazas, its acequias, the houses have their jalousies, and the stranger never loses a foreign feeling while he stays. It is true that there are large numbers of Germans, French, and Poles here, that no shop-keeper employs a clerk who can not deal with at least two of these nationalities besides his own, and the place is in a manner cosmopolitan ; but Spain is at the foundation of the whole of it. The secular buildings are such as those which the earthquakes had forced on the Spaniard in Mexico, and which, from habit, he brought with him—and wherever the modern builder varies the design, he ornaments the galleries with a light woodwork, cut, doubtless unconsciously, in a Moorish pattern—and the church buildings are such as those which the Spaniard venerated in his mother-land. The Cathedral of San Fernando has, indeed, been rebuilt, retaining only a small fragment of the old building at the back ; but the other ancient church buildings, quainter and more picturesque, known as missions, although in ruins, have endured no alteration of design.

San Antonio was itself a mission. A poor little village called San Fernandez in 1698, it was deemed best to remove thither from the Rio Grande the mission of San Antonio de Valero, in execution of a plan still further to settle and civilize Texas, and thus to repress the encroachments of the French, who, under the pretensions of La Salle's brief occupancy, were always laying claim to it. Thenceforth the mission was known as that of San Antonio de Bexar, from the name of the province, Bexar being an immense section of territory then comprising nearly all of southwestern Texas, attached to the Intendancy of San Luis Potosi. The population of the town was increased by a royal importation of families from the Canary Islands and from Tlaxcala, and during the following half century the missions of La Purisima Concepcion, of San Jose, San Juan and La Espada were built down the river, each a few miles from the other, and the Alamo was begun on the left bank just behind the town. These were posts partly religious, partly defensive, founded by the Franciscans, to whom some five square leagues were given for the purpose, and who induced the milder Indians to cultivate the rich lands, improve their own condition, and enlarge the revenues of the Church, without any doubt performing a great work of civilization. The buildings of the missions usually consisted of a noble church at one end of the square, a fort at the other, the apartments of the friars, the huts of the laborers, the granaries and storehouses distributed between, all of massive stone, and inclosed behind a high wall completing the whole as a fortress, which was, indeed, necessary, subject as it was to the incursions of the fierce northern Indians.

These missions have an interest for us quite apart from their beauty, for they stand up in their solitude and decay, still giving silent testimony to the immense debt that we, as a people, owe to-day to the old conquistadores of Spain. They are a part of the visible romance of our country, too ; for they met the line of that chain of forts which followed in the adventurous path of the Sieur de la Salle and the intrepid Father Hennepin from the Great Lakes to the Red River, and they also were the outposts of civilization in the wilderness. The monks of these missions, moreover, were those who opened to the world the resources of this great empire of the West ; with their patience and labor, they were the first pioneers of the region, and but for the riches which the soil displayed at their touch, the colonist might not have been tempted here for a century later. They cleared the way for a new power among the peoples of the earth, and in the annexation of that power to our own, in the war that followed and the consequent acquisition of all the northern half of Mexican territory and the great train of circumstances resulting, one sees that, like all the other conscientious workers of the world, they ' builded better than they knew.'

Every one of these missions is now a ruin ; the grass grows on so much of the roof as is left, the mesquite starts up in the long cloisters where the fathers used to pace, the cactus sprouts and blossoms in the crannies of the outer wall, the wild thyme hangs in bunches there, and sweetens all the lonesome summer air. Nothing can describe the solitary grandeur and beauty of the Concepcion, and the marvelous piece of color that it makes, as you drive over the prairie, first approaching it when, a mile and a half from the town, its twin towers and dome darkly rise on the luminous sky. It is the first religious ruin you have seen in America—indeed, these ruins are, we think, the only things of the sort in the country; its existence is a romance, its condition a mystery, and a vague pathos haunts its broken arches and disused cells. The mission of San Jose, some four miles below the first, is, however, both finer and more interesting. This is really, it is said, the mission of San Juan, but through a transmutation of names peculiar to Texas, in which, for instance, the original Brazos became the Colorado, and the Colorado the Brazos, the place is now always known as San Jose. The buildings of this, the second mission, were not only of finer design and workmanship, but they were those of a scholastic as well as of a religious institution, inclosed a much larger space, and are left in much more detail. The church was built in the style introduced in Europe by the Jesuits when the Renaissance had become wearying—the style from which subsequently the Louis Quatorze developed itself. But although a meretricious style, its effect, judging from these ruins, must have been very fine, particularly in the dazzling light of this latitude, and the execution of its details was of the best. The stone, although now lichen-eaten and weather-stained, is the soft cream-colored stone of the district, which is easily wrought, the surface walls frescoed with a diaper of vermilion and blue, of which only faint lines remain. All the lofty facade is a mass of superb sculpture of colossal figures, with cherubs, scrolls and flowers; similar noble work surrounds one of the exquisitely beautiful windows; but for the rest, the great halls are roofless, the long arcades are crumbling into mounds of dust, and even the fine statuary has been defaced by wanton wretches who have enriched themselves with the hand of a St. Joseph or the head of an infant Jesus. Such as the carving is, it is regarded with superstitious idolatry by the simple Mexicans whose village surrounds the ruin, and the priesthood itself would not dare to take any measures for its preservation that should remove it from their daily sight. The chapel attached to the mission is still in use, a weekly service being held there. In spite of its pretty font and of the groined arches of its vaulted roof, it is a sad spot. Two or three old paintings adorn it, a sacred image stands in the lofty niche of the only window, which, lined with scarlet, surrounds the image in a blazing aureole, while the walls all about the altars are strung with the votive offerings of the poor worshipers, who, since they can not give lace and jewels, and gold-wrought altar cloths, give curious patch-work hangings which are inexpressibly touching to see. There are said to be great underground chambers attached to this mission building, capable of holding two years' provision of wheat, together with

secret passages to the river, so that the water supply could never be cut off; and owing to this, the mission was able once to endure triumphantly, according to tradition, a siege of eighteen months' duration from those warlike Indians who never ceased their hostility to the undertaking of the Spaniard and the Franciscan. Of the other missions, down the river, there is scarcely enough left to mention; but take them by moonlight, the effulgent moonlight of San Antonio, and they are worth a journey to see, the front of La Espada towering above the dark foliage, a melancholy haunt of poetry and dreams. Why all these buildings have been allowed to fall into such a condition it is not easy to say. Whether it was that the secularization of the missions crippled them beyond their strength, whether the Indian service was no longer able to maintain them, whether the dry climate had any particularly injurious effect upon them, whether

the depredations of marauders have been equal to such destruction, or whether it is judged that they are most effective as they are—whatever the reason, the lover of the picturesque may well be thankful for the result.

The drive to these missions, in deep woods, across all the fords at all the windings of the rivers, through the forsaken avenues of pecans that the good friars planted, and up the open prairie-side, is as wide and delightful a contrast with the ruins as it is possible to imagine, and accents a great deal of their charm. Here is no decay, no disrepair. Nature is alive and throbbing through every leaf and blade; the mesquite is waving all light and feathery grace on every ripple of the air, a thing of beauty, half sunshine and half verdure; the mustang grape, with a stem the size of a baby's

waist, twists itself in long, loose ropes and
testoons from tree to tree, and spices the
wood ; the great ratamma, with its yellow
primrose flower set in a radiation of slender
green spike-like leaves, shines with all its
lamps against the dark masses of the magnificent pecan ; and earlier in the year the
wisache, each spray of which, strung with
downy golden balls, is precious in the Northern conservatory, soars like an illumination
beside the way, and the thickets of the lovely
frijo-lio clothe themselves in purple with the
narcissus at their feet.   All around the Concepcion mission, where one of the deadliest
fights of the Texan revolution once reddened
the grass, sheets of the white prickly-poppy
wave in the wind, the Texas star sprinkles
the sod, and the delicate little white rain-
lilies that spring after a shower, scatter their
delicious odor ; everywhere over the broad
slopes the prickly-pear blazes up in the sun
with its big red and yellow cups full of flame :
in the same colors, native to the soil, scarlet
and orange lantanas and abutilons grow beside the slender swaying mountain heliotrope
with its white blossom and vanilla scent, while
violet and verbena, morning-glory, marandia,
convolvulus, and clematis, greet the familiar
eye, and unknown blossoms flaunt in every
copse.   The music of the mocking-bird,
which tilts on countless topmost boughs, is
pouring over you in floods of ecstacy ; the
cardinal-bird's note pipes clear as he darts
from the shadow of one bush to another like a
winged coal of fire ; the little finches warble
and thrill ; the turtle doves coo on the low
boughs, or go skipping together across the
grass ; the scissors-tail and the chacalaca skim
over the tops of the thorny chaparral ; a flock
of blackbirds that seem to have lit on the knoll
in a patch of yellow blossoms fly away at your
approach, and take the blossoms with them ;
the rabbits bound along the ground ; the
splendid wings of butterflies brush your face.
Just below the second mission you come to
the falls of the San Antonio. Although the
falls themselves, divided into many, are of no
great height, yet the volume of tumbling foam,
the wondrous color of the waters, and all the
harmony of the world of verdure that in every
shade of mighty oak, dipping willow, and
feathery fern swings over the stream which
slips so smoothly to the fall, and with such
jewel-like polish that its very swiftness seems
stillness, make a picture of green and silver
that it would take a West Indian wilderness
to rival.
The Alamo, the last of the missions, and
one never quite completed, is but a few steps
from your inn, on a dusty plaza that is a reproach to all San Antonio. Its wall is over-
thrown and removed, its dormitories are piled
with military stores, its battle-scarred front has
been revamped and repainted, and market
carts roll to and fro on the spot where the
flames ascended at the touch of the torch of
an insolent foe over the funeral pyre of heroes.

But yet the Texan visits it as a shrine, and
thrills with pride in a history that is more to
him than all the Monmouths and Lexingtons
and Cowpens and Yorktowns of the Revolution ; for, after all, Texas is a domain by itself, with a past of its own, and although long
a voluntary member of our federation, yet,
like Hungary or like Scotland, it is hardly to
be absorbed.
The sword years since usurped the gown in
men's thoughts when they spoke of the church
of the Alamo, that fortress of the church mili-
tant. Yet many a stout contest, to be sure,
was waged in and around the little town of
Bexar before the walls of the Alamo were
ready for the banner poles from which such
various flags have tossed defiance : to-day the
French, under St. Denis and La Harpe, driving back into it all the Spaniards of the out-
lying country, to-morrow the Comanches and
the Tahuacanos harrying it, and even after it
was garrisoned, the Apache riding boldly in and
bidding the soldier there tether his horses.
But with the building of the Alamo the strug-
gles for its possession became fierce and frequent, and all the peaceful nestling beauty of
the town was, until within the last thirty years,
only the background for successive scenes of
bloodshed. Now Salcedo and Herrera surrender it to the Americans—that Salcedo
whose keen insight saw the ruin of Spain in
her colonies, and would have forbidden the
birds to fly across our border and bring back
any whisper of liberty ; now Elisondo threat-
ens it, one sunrise. from the distant heights of
the Alazan ; out of it eagerly marches a band to
meet Arredondo at the Medina, and lay their
bones to bleach on the old San Antonio road ;
now, again, a raw army of 500 men hold
Perfecto de Cos, the brother-in-law of Santa
Anna, and his force of nearly three times their
number, prisoners within the walls for two
months, till the assault is ordered, when, while
one party divert attention by an attack on the
Alamo, from which, as well as from the cathedral, waves the merciless black and red flag,
two columns march up Soledad and Acequia
Streets, the one pushing through De la Garza's
house, the other through Veramendi's—each
house, with its walls three and four feet in
thickness, being a little fort—push slowly on
day by day through the houses, not through
the streets, which were raked by Mexican guns,
through Navarro's house, into the priest's
house, into the square, when the black and red
flags come down and a white one goes up—a
bitterly contested fight, where on the second
day the magnificent Milan fell, long lying
buried where he fell. Although, some years
afterward, the ashes of this hero were removed
to a cemetery, yet the scarcity of land in Texas
recently created the necessity of running a
highway through the cemetery ; and while he
has his monument in the Capitol, yet Milan,
who so loved Liberty for Liberty's sake—lay
in canebrakes, slept in dungeons, starved and
bled and died for her—lies to-day in an

unmarked grave where every hoof insults him. But the great fight of the Alamo, that which has immortalized it with the battles of the world, took place when Santa Anna advanced upon it with all the machinery of war at Mexico's command. From the outset there was no hope within the walls, and the little garrison there made up their minds to their fate; indeed, one of them, Colonel Bonham, sent out to seek re-enforcements, came back alone, although he knew it was to die, heroically as Regulus returned to Carthage. There were 144 men in the Alamo; Santa Anna's troops, at first estimated at 1500, were presently increased to 4000; they were the flower of the Mexican soldiery, commanded by officers of matchless skill and daring, many of whom loathed the work required of them. But Santa Anna, who styled himself the Napoleon of the West, left no rise behind him: his policy was the policy of extermination. The town of San Antonio was already his; the blood-red flag flapped from the cathedral, and the fortress was summoned to surrender and throw itself upon Mexican mercy. What that mercy was can be imagined from the subsequent fate of those who capitulated with the brave, impetuous Fannin at Goliad under all the forms and articles of war, and with promise of speedy release, only to receive orders, one Sunday when they were singing 'Sweet Home,' to march out in double file under guard, suddenly halted when half a mile from the fort, the guard wheeling and firing upon them till they fell, betrayed and butchered in cold blood. 'This day, Palm Sunday,' writes a Mexican officer of the massacre, 'has been to me a day of most heart-felt sorrow. At six in the morning the execution of 412 American prisoners was commenced, and continued till eight, when the last of the number was shot. At eleven commenced the operation of burning their bodies. But what an awful scene did the field present, when the prisoners were executed and fell dead in heaps, and what spectator could view it without horror! They were all young, the oldest not more than thirty, and of fine florid complexions. When the unfortunate youths were brought to the place of death, their lamentations, and the appeals which they uttered to Heaven in their own language, with extended arms, kneeling or prostrate on the earth, were such as might have caused the very stones to cry out in compassion.'

Travis and his men had no mind for such mercy. Shut up in the Alamo, this was the proclamation of that superb leader: 'I am determined to sustain myself as long as possible, and die like a soldier who never forgets what is due to his own honor and that of his country. Victory or death!'

This splendid death-cry was unheard. The call was neglected. No help came. Santa Anna surrounded the place on all sides with intrenched encampments, and kept up a cannonade for ten days, many times attempting to scale the walls, but always repulsed with slaughter—1500 of his men, it is said, falling before the unerring Texas rifle. At midnight of the thirteenth day the storming party was ordered to the assault for the last time, the reluctant infantry, pricked on by cavalry in the rear, amidst the roar of artillery and the volleys of musketry, the trumpet sounding the dreadful notes of the *dequelo*, signifying no quarter. Twice they made the attempt in vain, and recoiled only to be urged on for the third time by the irresistible cordon behind them; the third time they mounted the walls and fell to their bloody work. It was short and terrible. As Travis stood on an angle of the northern wall, cheering the fearless spirits behind him, a ball struck his forehead, and he fell; a Mexican officer rushed forward to dispatch him, but he died on the point of Travis's sword as that hero breathed his last. And with that the indiscriminate slaughter began, man to man, of the little force that, worn out

with the task of repelling attacks and manning works that required five times their number, with sleeplessness and thirst, and without time to reload their pieces, fought with their knives and the stocks of their rifles till no soul of the desperate band was left alive. Death and Santa Anna held the place. The Alcalde of San Antonio, summoned before the conqueror, pointed out to him Travis on the wall with the bullet in his forehead, Bowie butchered in his cell where he lay on his sick-bed, Evans shot in the act of blowing up the magazine, and David Crockett lying dead with a circle of his slaughtered foes around him. On the shaft erected to the heroes runs a legend whose eloquence makes the heart stand still: 'Thermopylæ had its messenger of defeat, the Alamo had none.'

San Antonio has always had more or less to do with warlike operations. It is now again a military department, under the control of

General Ord, a brave and accomplished soldier, who from this station directs the frontier movements between those forts that constitute our wall against the Mexican and the Apache ; and in the event of a new Mexican war, it will be, as it is now, owing both to its situation and to its railway connection with the coast, the base of military operations. It has an arsenal, with picturesque grounds and buildings, on Flores Street, and a military depot on one of the side hills, whose stone walls inclose sufficient accommodation for all the stores needed in time of war, while its tower overlooks the country for many miles. There are several regiments permanently stationed here, while officers of other regiments are frequently going and coming to and from other posts ; and there is almost a pathetic contrast between the young officer with his unfleshed sword, who arrives smooth and fresh and fine in San Antonio, and the bronzed and roughened fellow who rides back from the frontier after a couple of years of service there. San Antonio is held to be quite a desirable post in the army, and the army life adds a great deal to the pleasure of society in the place, with the high tone of its brilliant men and lovely women of varied experience and graceful manners. But the society proper to the place itself is of a superior order, having something of the old Spanish base of courtesy and gravity with the polish born of contact with the world. For the San Antonians are by no means a stay-at-home people, nor do they confine their rambles to Mexico and the South ; you will find many of those in comfortable circumstances who have made the European tour, and several who have crossed the ocean half a dozen times. Besides the school of the convent, there are several fine private schools, and there has long been a system of free schools in operation there, and those for whom these facilities are insufficient send their children sometimes to the North and sometimes to Europe. Of the young ladies there—who, by-the-way, are rather remarkable for their beauty—there are many who speak Spanish or German, and many are mistresses of four tongues ; while several of the matrons have an acquaintance with the dead languages which would allow them to fit their own boys for college, are well read in general literature too, and proud of the fact that Texas has no lack of literature of her own. A good deal of the quality of this society is owing to the fact that its members depend so largely for entertainment upon themselves, and while dancing and music have received great attention, the art of conversation has had an unconscious cultivation that it does not so generally receive where opera and concert and theatre spare the trouble. Yet this society is a growth of the present century. When the first American lady went to San Antonio, the Mexican women would beg permission to come in and admire her, and after sitting in silence a space, would go away lisping many thanks in their sweet syllables, and saying

that she was very white and very lovely. This soft lisp of the Mexican, we may say, has somewhat infected the speech of the average San Antonian, who calls acequias isakers, and speaks of the Salado and the Cibolo as the Slough and the Seewiller. Perhaps, also, he has been infected by something more than the Mexican lisp, in a certain enervation and lack of public spirit which causes him to allow his lovely town still to retain its fantastic charm, instead of joining the march of improvement ; he does not wish to see things other than they always have been ; it is no paradox for him to say that although they be better, they are not so good. This is the square where Baron de Bastrop met Moses Austin turning away in disgrace and despair, and changed the fortunes of Texas ; here is the public crossing of the river where old Delgado's head was set up on a pole ; there is the brook that once ran red with the blood of Salcedo, Herrera, and twelve other good knights and true to pay for that head ; and yonder is the plaza where the famous Comanche fight took place not forty years ago, when three-score Indian warriors, squaws and papooses came into San Antonio by appointment to surrender their white prisoners, and, failing to keep faith, were told they should be held as hostages, upon which, in an instant, bows were strong and knives unsheathed, and in the fearful struggle which followed, the squaws themselves fighting like tiger-cats, not one of the warriors was left alive. The old San Antonian wishes to keep these places unaltered, nor would he have the honored names of Manchaca, Navarro, Zavala, Seguin, and their sort, superseded by those of enterprising emigrants. From the point of view of the picturesque he is certainly right ; but otherwise one is reminded of the saying, now become a proverb, that the enemy of Texas is the old Texan. In spite of him, though, certain changes will be wrought by time ; enterprise has already crept into the place. It has a Historical Society and a Board of Trade ; it is talking of a new system of sewerage ; it has a gas house, much of whose gas is made of cotton-seed ; it has a railroad which has already improved its market, and it is bound to have others yet.

There is a sort of romance attaching to the road that brings into daily communication with the world this city, one of that lonely trio, San Antonio, Nacogdoches, and Santa Fe, that for nearly two centuries have stood on their long untraveled trails, unknown and remote in their silent solitudes upon the outposts. This road was built, single-handed, by its owner, Mr. Peirce, who is said to be the largest land-holder in the world. The bed in all its length is broad and firm, much of it made of the solid concrete deposits which are found on the line, the ties are laid with an exact precision, the rails are steel, and the bridges are of iron, with piers of solid masonry that defy the floods. On the occasion of its opening the San Antonians displayed a unique hospitality. To every guest that came over the road they gave literally the freedom of the city—the best they had to offer. Bed and board and fruit and flower were his; any garden where he wished to stroll was his; any carriage that he chose to stop upon the street and enter was his; any bar across which he wished to drink—and their name is legion— any cigar he chose to take. For three days the three hundred guests were entertained as kings and princes entertain, and were dismissed without having been allowed to pay a bill. It has always been a long and fatiguing stage-coach ride thither; but now the Texan is pouring in to visit its sanctuaries. He calls it almost invariably 'Santone,' and it is as full of novelty and delight to him as to the rest of the world. He goes to the Alamo and is weighed, congratulating himself on those that were weighed in the balance and not found wanting there once before; he climbs to the top of the mission tower, and recalls yet earlier days; he visits the springs; and he spends his evening at Wolfram's Garden, where the cups of colored light, among all the greenery reflected in the river, make an elfin place of strange contrast to those rude earlier scenes. He goes to the bee caves outside the city, to the bat caves some twenty miles away, where the scent of ammonia is stifling, the accumulations of guano are tremendous, and where the bats flying out just at sunset in long streams, like the never-ceasing smoke of a volcano, darken all the air, while the transparent membranes of their outstretched wings, catching the sidelong sunlight, make an unintermitting dazzle of prismatic lustre. Or perhaps he is on the fortunate party that unearths the skeleton in armor of one of those Spanish knights sent out by Cortez to find the seven treasure cities and never returning—wonderful bronze armor, finished in the perfection of art. Within the town he sees the long emigrant train threading the streets, with homesick women and determined men; he sees the great supply trains going out full and heavy to the markets of Saltillo, Monterey, and Chihuahua, and returning with hides and silver and wool; he sees the hunters coming home laden with game, and the gay party of young roughs pushing forth, with their six-shooters on the saddle, to seek for the lost mines of San Saba, or for those of Uvalde and the remoter west. He sees, too, the group of Mexican officers meeting here, perhaps for refuge, perhaps for safer conspiracy, perhaps to act with that Escobedo who put an end to Maximilian's pretty romance, and served notice on Europe to send no more kings to America; he sees the old banker, who, an American prisoner, has cleaned the streets of Mexico with ball and chain on his foot, the old physician who holds the diploma of all the learned societies in Europe, and who came to this country with that scion of royalty, the prince who colonized New Braunfels, bringing with him letters from Humboldt; or possibly he may meet a still stately dame who wears the diamonds given to her by her old partner in the dance, the pirate Lafitte, hero of Byron's 'Corsair.' He sees, with these, this and that veteran of Houston's men, still full of the old fire, as interesting to him, and almost as ancient, as if just stepped out of Joshua's army before Jericho; or, as possibly, one of the 'bean men,' a sort of sacred character, being the survivor of the famous Mier expedition.

But her own surrounding hills and prairies are wealth enough for her as it is. The yield of the cereals there is simply enormous. The corn is twice as high as your head in May, and the grass has twice been cut by that; every known vegetable has long been in the market then. The sweet and luscious figs are ripe, and pears and apples, apricots, plums and

peaches, are ready to gather; while, later in the year, bananas, pomegranates and persimmons come in, and the pecans drop big and sweet as one finds them nowhere else. There are fields about San Antonio where $400 an acre have been realized out of sugar-cane, although that is an extremely exceptional yield, the proceeds being partly due to the sale of cane in the streets, it being a choice morsel in its season. Large quantities of it are fed to cattle also; and for them, as another delicacy, the prickly-pear is raked into heaps and scorched of its thorns by fire. The Spanish moss is found in immense quantities on the trees in certain portions of country round San Antonio, as well as all the way to the coast. It is an epiphyte, not a parasite, drawing its sustenance from the air, and not the tree, to which it does no injury; and it is already forming a good branch of commerce, as, being well rotted and dried, it makes a valuable substitute for curled hair in upholstery. Cotton, too, is almost equally prolific with everything else. In fact, there is nothing which the rich earth does not seem capable of producing, and producing at its best. As you see it freshly turned up, clean, dark, and glistening as though it held hidden sunbeams, it seems, according to the old saying, fairly good enough to eat. It would excuse the clay-eaters themselves if it were on such substance that they fed;

and one would well wish that, having the tra-
ditional peck of dirt to eat, it might be eaten in
San Antonio.   One does not wonder to see this
sod break into blossom the day after it is cut.

> ' A footfall there
> Suffices to upturn to the warm air
> Half-germinating spices: mere decay
> Produces richer life, and day by day
> New pollen on the lily petal grows,
> And still more labyrinthine buds the rose.'

And San Antonio in this matter is but the type
of all western Texas—a land of promise and
of plenty ; a land flowing in milk and honey
(if, with the cattle roaming in multitudes, one
were not obliged to use condensed milk in
one's coffee) ; a land where the vagrant can
sleep in comfort under a tent in open air all
his lifetime, and may live in luxury, scarcely
lifting his hands to labor, and where the energetic and intelligent bind fortune hand and
foot, and compel her to their service. Nearly
300,000 people entered it in the last year, and
sought permanent homes ; many more, we
understand, contemplate the same movement
in the coming year.   And their success is
entirely in the measure of their endeavor ; for
with eggs selling at from six to ten cents a
dozen, and beef at from five to eight cents a
pound, the cost of living is at its minimum.
Rents are the only expensive item, and the
climate, as we have said, makes a tent sufficient shelter until a house can be built. And
never was any place more full of opportunity
to those who can seize occasion by the fore-
lock—opportunity, too, quite outside of the
farming industries.   Wonderful water-powers
that could spin and weave all the cotton on
earth compass the cotton belt there, while the
machinery of woolen mills could run without
steam beside the ranch where the wool is
shorn ; the huge heaps of bones, gathered
from the prairies where the cattle of two hundred years have laid them, and that are
transported at great cost, could be ground
into dust, or made into combs and buttons on
the spot ; acres of blooming wild white pop-
pies tell what is yet to be done there in opium;
tons of indigo are ready to the hand ; and the
mesquite is able to tan the hides that travel
some five thousand miles before they come
back in saddles and harnesses and shoes.

This mesquite, by the way, could be to the
Texan almost as much as the palm is to the
Arab—an object of pleasure to the eye of
man. Cattle browse upon its foliage, sheep
eagerly eat its beans ; its gnarled wood, when
grown to any size, is as fine as old mahogany
for furniture ; its abundant gum is the gum-
arabic of the East, and its bark tans leather
as quickly and thoroughly as any other sub-
stance known.   Forbidden by Spain, in that
narrow policy which has reacted in ruin on
herself, to grow flax, hemp, saffron, olives,
grapes and mulberries, the country blossoms
with them all to-day.   And, in truth, there is
nothing which she does not bring forth, from
the wines of El Paso to the camels raised and
sold to traveling menageries, for confiding
parents to exhibit to marveling children as the
ship of the desert, and the product of the
Scriptural East.

It is the Scriptural East that the region
round about San Antonio, and all this western Texas, indeed, constantly presents to
the mind in the lay of the land and all its
characteristics.   The irrigating ditches, the
shepherds and their flocks, the cattle on a
thousand hills, the wild asses snuffing the
breeze, the wheat, the vineyard, the lilies
of the field, the smell of the grape, the voice
of the turtle-dove, the fig and the pomegranate
—they are all there ; the very atmosphere,
and the high clear heavens recall the skies
of Palestine ;   one feels what the burden
and heat of the day means, and recalls the
Lord walking in His garden in the cool of
the evening.   At every step some memory or association concerning the Holy Land
arises ; and the dweller, sitting on his gallery, and overlooking his green pastures,
as the sweet and sudden dusk follows sunset
without twilight there, can well give thanks,
saying, ' For the Lord thy God bringeth
thee into a good land, a land of brooks of water, of fountains and depths that spring out of
valleys and hills ; a land of wheat, and barley,
and vines, and fig-trees, and pomegranates ;
a land of oil olive, and honey ; a land wherein
thou shalt eat bread without scarceness, thou
shalt not lack anything in it ; a land whose
stones are iron, and out of whose hills thou
mayest dig brass.' "

SAN ANTONIO OF TO-DAY.

AS the tourist or invalid, perhaps, desires some description of San Antonio's new features, and, may be, likes to know also something of rambles of former times in a stage-coach or in a prairie schooner, we will give some further synopsis of such rambles made very recently, and of such made in former years.

A stage ride from San Antonio to Austin was considered one of the most romantic in western Texas, and in fact, by being favored with a comfortable seat beside the driver, it was a rare treat. We will, in our limited space, only speak of a drive from San Antonio to the Comal Springs, near New Braunfels.

One beautiful morning we take our place beside the driver of the San Antonio stage, on the high box-seat, perched above four fine and strong horses, in front of the Menger House. The driver cracks his long whip, and we dash by quaint stone houses, pass the Government Depot, climb a hill, and get some glimpses of the dear old city, encircled by the richly tree-bordered river ; in the dim distance some missions peep out between pecan groves ; to the left we observe a long chain of the Guadalupe Mountains, and are on the hard, white, glistening highway.

Fording the Saluda Creek, we come upon something of a plateau, overgrown with nutritious pasturage grass and shaded by the scant foliage of numberless mesquite orchards, that crowd each other on either side of the road. Here a dilapidated Mexican hut varies the scene, and further on a beautiful farm-house, with its surroundings indicative of thrift and prosperity, pleases the eye ; anon a prairie, whereon graze the cattle in innumerable herds, and where the frolicsome calf disports himself with his playmate in the presence of a sleepy mother ruminating upon the pickings of the past feed time. In the distance a couple of cow-boys—one with his Mexican sombrero and flashy ornaments, the other with slouch hat and boot-legs over his pants—pursue a stray member of the herd and bring it within bounds.

Overhead the bright blue sky reaches from horizon to horizon, and the pure atmosphere all around fills the lungs at every inhalation with the essence of eternal youth. Onward we whirl, passing low stone fences that inclose cultivated fields and growing crops ; these marks of the husbandman indicate the prox-imity of a streamlet, on the banks of which nature has planted a more fertile soil. In its crystal depths our horses slake their thirst,

and we ascend again a still higher elevation of table-land. From here, as by magic, the beautiful little city of New Braunfels, nestled in the bosom of the rich valley of the Comal River, bursts upon our view. As we draw near, the increased number of little homes betoken the vicinity of a manufacturing town, and a growing industry among the settlers.

For a moment we lose sight of the city and are buried in the dense shades of a live-oak, where the great branches of the trees reach out their arms seemingly to support the long

rolls of hanging moss, almost sweeping the ground as the breeze stirs among green leaves. Next, on the left, the white rock foundations, in contrast to a rich growth of green cedars, make a beautiful picture of wild nature's painting. To the right the hand of man paints a rural picture with all the colors of incipient, progressing and perfected agriculture.

The yard, the garden, the field are all there tilled by the careful hand and hoe, or by the brawny muscle and sharp ploughshare A fair maiden with flaxen hair and bright blue eyes looks up from her work among the rose bushes as we pass, and seems not only queen of that domain, but queen of health, beauty and happiness. We imagined she almost smiled as we passed on to the town of New Braunfels, the oldest settlement in that section, it having been the home of German settlers as early as 1854. There they set foot, and despite the hardships of pioneer life, privations of explorers and frequent battlings with Indians, held, and will ever hold, a firm footing on that now prospering land. Having safely landed in town, we met at the hotel Mr. Landa, whom we found to be an elegant, educated and refined gentleman and the fortunate owner of Comal Springs. Resting awhile we accepted an invitation from him to visit these springs. Seated in an open landau behind a fine and spirited pair of dark bays, we drive through the principal streets and see on every hand

4

groups of healthy looking children indulging in their more or less athletic sports. Now and then the graceful movement and elastic step of a fair-haired German lassie attracts the eye, while the clear ring of her sweet voice speaks to the ear in prophetic tones of the generations to come and the great people that are yet to cover the lands of this great state. The town is laid out with much regularity, giving each house ample room for a vegetable or flower garden. The Germans having a peculiar fondness for flowers did not forget to leave space for the cultivation of roses, pinks, resedas, violets, forget-me-nots and the many other fragrant flowers that flourish here. Driving from the city we descend a small decline and cross one of the many tributaries of the Comal. To the right a miniature Niagara tumbling over the rocks conveys to your mind a small idea of the immense water power confined within the banks of this river. Ascending we pass Landa's grist mill, built of stone, just above which the waters, pent up by the dam, form a beautiful lakelet of crystal-clear water. Around its borders, and hanging over its limpid depths, water lilies, the giant leaves of the caladium and many other aquatic plants are to be seen in the wildest profusion and of the most luxuriant growth. Here we dismount, and on a rustic foot-bridge made of unsawn forest saplings, cross to an island whereon is a vineyard of the most luscious grapes to be found anywhere. Side by side with the vines, are fig-trees loaded down with an abundant crop of that most delicious fruit.

From this garden-island, from this lap of luxury, we take our departure in a small canoe and skim over the placid surface of the Comal, with rich wreaths of aquatic plants and tropical flowers clustering along either margin and the dense foliage of giant forest trees clasping hands above the silvery band that steals its way about their roots, until we re-enter the carriage to drive past spring after spring of bubbling water, throwing out its share of the immense supply that forms this beautiful river of which we write. As we travel there stretches behind us one of the most beautiful and sublime landscapes that the pen of poet or novelist ever described, or the brush of painter pictured. Not the celebrated vale of Tempe, nor any of the immortal valleys of the Orient equal this paradise of the western world. After having ridden some distance we again take it afoot and cross once more to our little insular Eden, from whence in another direction is seen the loveliest picture of nature that enchanted the human eye. The undisturbed waters stretch away as far as the eye can reach, and in their shallows the white cranes wade and the moss dips to the water's surface from the overreaching boughs. A little way off the dark green cedars spring up from the white limestone rocks, and all are reflected from the glassy mirror that trails beneath them. We were fain to exclaim

"Let us rest here forever!" Through a picturesque glen we again find our carriage and drive to the top of some mountainous bluffs where begins a table-land more beautiful than that seen on the road from San Antonio. It is covered with rich grass, and numberless clumps of trees are interspersed with meadows replete with the rarest flowers. In the distance the blue mountain ridges loom up, deepening their tints as they recede, and valley after valley intervenes. Evening has now nearly gone ; and as the sun, casting his farewell gleams into every nook and corner of the landscape, and fringing every cloud with his silvery lining, disappears behind the western horizon, a scene of grandeur is presented that beggars all description.

San Antonio is watered by two beautiful streams, the San Antonio and the San Pedro, the former running directly through the heart of the city. It is a crystal-clear, swift current, of a deep bluish hue ; and although flowing in a channel not very wide, its banks are picturesque and overhung by beautiful foliage that in some places (often in the middle of the town) overarch the entire stream. Passing under large stone bridges and under newly built iron bridges, it is one of the loveliest possible water streets, and is something original and unique, to be found nowhere else. The flights of stone steps, leading to cozy bath-houses and thence into the comfortable villas, reminds one of Venetian canals. Enchanting gardens touch the borders of this river, and little skiffs are used in making neighborly visits. The residences on Flores Street are all completely embowered in shrubbery. Since 1877 many additions have been made to the building list of 1876, chiefly on Main Street ; they are notable for the decided architectural improvement and the superiority of the building material used.

The headquarters of the general commanding the Department of Texas are located in a large stone building two stories in height, on Government Hill, near the new Government depot, which is situated upon a splendid elevation northeast of the city, with a beautiful view of the town, commanding both banks of the San Antonio, with all the missions, and the range of hills semicircling the city with a gradual elevation of 200 to 300 feet. All the grounds comprise 216 acres, with four main buildings constructed of stone, each 624 feet in length. The offices of the different departments are perfect and complete in every respect, and are very elegantly furnished. The building to the west is for storage of all kinds ; that upon the east for flour ; upon the north for wagon, blacksmith and repair shops, etc. In the centre is a tower built of stone, 90 feet in height, a landmark, and one that commands a panorama rarely to be seen. To the northwest are corrals, with buildings for all domestic purposes, and in regard to combining the useful and convenient no other military post in the state is its equal. Plazas, parks,

MISSION SAN JOSE.

churches, missions, etc., are already described. In regard to business houses, information can be found in the newspapers of this city, which are ably conducted and have a leading influence in this part of the state. The *Express* is published daily and weekly ; the *Herald*, also, is published both daily and weekly, and last but not least, the *Freie Presse*, a German paper, the very best of its kind in the state. The leading hotels are the Menger, Maverick, the Central House, Hord's Hotel, and the Vance House. Private boarding-houses with all accommodations can be easily secured. Places of amusement are few, but with the growth of the city, suitable investments will doubtless be made. At present only three halls are in use : the Alamo Literary Association, the Casino (German), Turner and the Music Hall above Schalz's summer garden.

The markets, mentioned elsewhere, present scenes unlike those in any other city in the United States, and remind the curious looker-on of the customs of an old Spanish town.

Concerning the public and private schools of San Antonio, we can state that they do great credit to the city, being amply provided for by the State School Fund, and receiving besides a liberal share of the Peabody and private endowments. Five public schools and buildings ; eight private and select schools ; one Catholic college ; one convent, and one English and German academy (the last being one of the best conducted schools in Texas), constitute San Antonio's educational advantages. The various associations are, Masonic, Odd Fellows, Knights of Pythias, and other benevolent orders ; a German casino, with a beautiful theatre and concert hall, a Turn Verein, an association of Mexican war veterans, an Agricultural, Stock and Industrial Association and two well-organized singing clubs (German). Carriages, street cars and other means of transportation are convenient for the many beautiful drives or excursions to the missions and parks near the city. Objects of interest can be found everywhere in this old town ; and the superior character and comfort of nearly an hundred licensed carriages, afford to the visitor and invalid a necessary luxury, which should be highly appreciated, as many other winter resorts are often lacking in such accommodations.

The water supply of San Antonio is the very best in the state. From the earliest settlement of the quaint town, the San Antonio and San Pedro springs and rivers have been the only sources of water supply for household and gardening purposes. The rapid growth of the city being foreseen, it was decided to erect a system of water-works, using the head-springs of the San Antonio River to supply the city with an abundance of pure water for fire protection, sanitary, public and domestic purposes ; these works are now completed. The head-force raises the water, which is of the purest kind, to a height of twenty-four feet, from whence it is forced by

machinery to the height of eighty-five feet into the receiving reservoir. This reservoir is erected upon a space of six acres, known as the City Rock Quarries, having a capacity of five million gallons. Not far from these waterworks are the sources of the San Antonio. In no other place is the river so supremely beautiful as at its head-waters, on the high plateau at the foot of the Guadalupe range ; here it breaks out from innumerable springs, which at once form themselves into a beautiful stream, having its banks ornamented with a wealth of blossoms. Mr. Brackenridge, who purchased the estate of several hundred acres containing this lovely natural park lying along the base of the mountains, has thrown a protecting stone wall around it to prevent abuses, but has kindly given one or two days each week to those who desire to visit these most interesting and picturesque grounds, the possession of which an English lord might envy him. As some one well says in regard to the San Antonio : "The stream is a delicious poem, written in water on the loveliest of river-beds, from which mosses, ferns, dreamiest green and painted crimson, rich opalescent and strong golden hues peep out. Every few rods there is a lovely waterscape—a painting in miniature." Butterflies of rare beauty and other brilliant winged insects, swarm from flower to flower, large-leaved bananas overshadow a part of the springs, and groups of other beautiful water-plants lend an enchanting verdure to the cool banks. Noble pecans, beautiful oaks, ashes and shrubs of every description, "stand like Druids of old," around these fairy-like waters, not only draped with the historic moss, but festooned with gala-vines, with blooms of red, white and blue, reflecting themselves in the placid pools so pure, so deep and crystal-clear to the very bottom. It is the desire of many prominent citizens to purchase this splendid park for the city, to be used by the public, and Mr. Brackenridge is willing to part with it for the sake of the great general benefit that will arise from its proper use. The San Pedro is also a clear and charming stream. All its sources are confined to the northern border of the city in the enclosure of San Pedro Park, easily reached by the street cars. This park is a favorite public resort ; there are held festivals, illuminations, boating parties on the lovely lakelets, picnics, concerts, hops, fairs, etc. The San Pedro is utilized to irrigate a large portion of the city, and from its very source becomes a great benefactor, as is its noble companion stream, the San Antonio. Within the corporate limits the San Pedro produces, by a system of irrigation, a luxuriant growth of trees and brilliant as well as fragrant flowers in the adjoining gardens. Tropical blossoms, and beautiful cypresses, weeping willows and other shade-trees with variously tinted foliage, delight the eye on every side, and mark with beauty the course of the silver stream.

SAN ANTONIO AS A WINTER RESORT.

SO much has been said and written over and over again in regard to San Antonio and its highlands, its unsurpassed climate, so healing and soothing to the invalid, that we can only recapitulate some general remarks by experienced observers, in order not to lengthen this pamphlet too much ; and we will bring it to a close with some general hints to invalids, as well as to the tourist or sportsman, and by giving some description of the Sunset Route's Pacific Extension and its adjacent winter resorts.

Thousands who have the means wherewith to travel —especially those residing in the cold climates of the North and East, and who wish to avoid the extreme winters, lasting about seven months of the year—and who are in search of a more genial clime, would do well to visit this section. Many valuable lives can thus be saved and many prolonged. Those suffering with pulmonary affections, rheumatism or kindred infirmities, will be cured or relieved by spending the winter months in San Antonio or its vicinity. The elevation of San Antonio is 625 feet above the sea. Its atmosphere is as pure and balmy as that of Italy ; its climate is dry and invigorating ; its numerous springs are pure and sparkling; summer flowers bloom in the gardens during the winter ; in and about the city are easy and delightful drives ; hotels and private residences offer all the conveniences and luxuries that are found anywhere, either North or East. The medical faculty, noted for their science and high attainments, are ever ready to help and to advise. It is an invariable experience of those who go to San Antonio from the heavy and moist atmospheres of the Mississippi valley, the marshy swamps of Florida and Cuba, or the Atlantic coast, that they experience a sense of

exhilaration and buoyancy while inhaling the strengthening, electrical and balmy atmosphere in this western Paradise, unequaled in climate throughout the United States. Those sceptical ones, who are desirous of being convinced of the fact that the climate of western Texas is equal in salubrity and healthfulness to

DOORWAY OF SAN JOSE MISSION.

any in the world, have only to consult the United States Census Reports of the Signal Service Observations, or any scientific work on climatology ; an honest perusal of these would certainly satisfy the most incredulous. We give the mean standard thermometer as observed for the six cold months: November 58.73, December 55.76, January 49.59, February 58.72, March 64.50 and April 67.67 degrees.

It is an interesting fact that the inhabitants of western Texas are free from pulmonary diseases ; a truth noticed even in its first settlement in the seventeenth century. No such diseases can originate in that section of the state, while many who go there suffering from lung affections, often experience a perfect cure or prolong their lives for years in comfort. Marked instances of this have become widely known, and the General Passenger Department will with pleasure answer any questions in regard to persons who have been thus almost miraculously cured, giving names and address of ladies and gentlemen of wealth and culture.

· It is further stated that invalids in an advanced stage of consumption will find almost any desired temperature near San Antonio, securing a beneficial change by going to higher altitudes, so easily reached by the extension of the Sunset Route. Such health-seekers will receive from the medical practitioners of San Antonio the most reliable and trustworthy advice.

Many an invalid in the first stages of lung disease would be cured by camp life, by sleeping in a tent, and pursuing out-door sports ; and many who make an experiment of frontier life soon delight in their wild freedom, and learn to enjoy a venison steak, a trout, a wild turkey, a partridge or prairie chicken, simply prepared by the camp fire, in preference to all the luxurious dishes by any "cordon bleu." The reward for such a hardy experiment is almost invariably a recovery of strength ; and the bloom of youth returns with the bronzed cheek and health with the muscular chest and clear eye. The following extracts from a paper read at the ninth annual session of the Texas State Medical Association, by Dr. J. B. Robertson, one of the oldest and most highly esteemed physicians in the state, will not fail to impress the reader favorably :

"That portion of western and southwestern Texas lying west of the 98th meridian of longitude, and north of the 29th degree of latitude, has an elevation above the sea, beginning fifty miles south of San Antonio, of 500 feet, and gradually rising as the line is traced north to 2000 feet. This region is drained by the following rivers and their numerous tributaries: Brazos, Colorado, Guadalupe, San Antonio, Nueces and Rio Grande, all of which find their outlets into the Gulf of Mexico. The rapidly decreasing elevation of the country through which these streams pass in their course to the sea, secures to the section named the most perfect and thorough drainage. In addition to this fact, this vast area of territory is entirely free from ponds, marshes, lakes or stagnant bodies of water, to disturb with their contaminating effluvia the purity of the atmosphere. Here are also found the principal mountain ranges, of which the Guadalupe is the largest, and has the greatest elevation. These mountains, with their intervening valleys and plains, with their springs of pure and limpid water—which, for beauty and picturesqueness are rarely equaled and never surpassed—are beginning to attract the attention of the professional man in search of a locality for the climatic treatment of diseases of respiratory organs, especially phthisis.

"It is a source of much regret that I have not been able to get satisfactory reports of the range of the thermometer and the humidity of the atmosphere. I am only able to give the mean temperature for the seasons and year (means obtained from six years' observation at San Antonio) ending with the year 1875, which is :

Spring, . . .	69.94 deg.
Summer, . . .	85.56 "
Autumn, . . .	68.95 "
Winter, . . .	52.94 "
For the whole year, .	68.85 "

"The pressure of vapor, its weight, the absolute humidity, have, as far as I know, never been measured, but the observations of daily life, by all who have lived in any part of this section or traveled through it, concur in attesting the astonishing rapidity with which the roads dry after a fall of rain; and the perfect preservation of meats for days, hanging in the open air, indicate unmistakably a small amount of moisture suspended in the air.

"The beneficial effects of the climate, in the area treated of, is not simply a matter of opinion, on the part of the writer, on purely theoretical grounds. During a practice of over thirty years in central Texas he has seen many patients sent there with clearly marked indications of consumption, and at a time in the history of the country when such patients had to rely almost entirely upon the climate for the benefit they received. In all cases the change gave marked relief, with, he believes, a prolongation of life for years with some and a perfect cure with others."

The desire for winter resorts has become so widespread throughout the North and East that the superiority of western Texas in regard to climate will not fail to attract attention.

MEDINA COUNTY.

Situated west of Bexar County, on the 22d degree of longitude and between the 29th and 30th parallels of latitude, this county has an area of 1304 square miles, and a population of about 5000. The surface of the county is a somewhat elevated and undulating prairie, in the region of the finest stock-raising country in Texas, and has, like Bexar County, perpetual pastures of the finest and most nutritious grasses, with an abundant supply of clear and running water, rendering it a magnificent grazing country, where stock thrive and fatten throughout the year. The Medina River, a branch of the San Antonio River (crossed by the railroad over a fine bridge), with high banks overhung by the foliage of majestic trees, has beautiful landscape scenery. This river takes its course through the eastern sec-

VIEW ON HEADWATERS OF THE SAN ANTONIO.

tion, while the Quihi, Chacon, Hondo, Verde, Geronimo, Ranchero and Black Seco Creeks flow through other sections of the county.

The soils are diversified and range in character from the rich bottom-land, well timbered along the margins of the streams, to the lighter soils of the uplands. About one-eighth of the county is timbered with the several varieties found in this section of the state, post-oak predominating, and upon the prairies the mesquite abounds.

Farmers produce the usual crops in this section—cotton, corn and the smaller grains, all kinds of vegetables and some fruits, especially the grape, which is abundant. Sheep husbandry is particularly profitable in this county, and this industry is increasing very rapidly. The Sunset Route, traversing the county, affords excellent facilities for transportation of surplus productions and live stock to the best markets in the country. The building of this road through the county has added largely to its wealth and population; and as the people are progressive, intelligent and hospitable they receive a large accession from the constant flow of immigration. Good lands can be bought at from $1 to $5 per acre upon favorable terms, and the advantages offered by this county to industrious laborers and farmers are many and hardly excelled in any other section of the state. There is also an abundance of fine stone for building purposes, which can be quarried at a moderate cost.

A railroad branch from the main line will soon reach Castroville, only four miles, the county seat. This town is situated in a beautiful valley twenty-five miles west from San Antonio. It was settled in 1844 by French and German immigrants under the direction of Henry Castro. The dwellings, hotels, churches and business places of Castroville are mainly constructed of stone. It has a sawmill, two grist-mills, Protestant and Catholic churches and free schools. As the elevation of this lovely city is over 500 feet, and other sections near by 700 feet above the level of the gulf, from which it receives the cooling breezes from the south, the climate is delightful and charming; and as the health of the place is unexcelled, the mean temperature being about 68 degrees, it will become a favorite winter resort for invalids. The tourist and hunter find also all accommodations for their rambles, and can readily obtain an invitation to accompany the hardy Texans upon their annual "rounding up," which occurs in the early summer, when all the cattle upon a range are brought together, the calves marked and branded and the selections made for market.

Traveling in west-bound trains one often notes some consumptive who is journeying toward the dry plains to rough it for a while, with hope and good prospect of regaining his health. Many others have sought these prairies, from the close confinement of mercantile life, to renew failing health and what medicine could not heal, and were cured by the pure atmosphere; and ranches are full of brown and bearded men, in full possession of strength, who came to this locality under such conditions as mentioned above. Since the Sunset Route traverses this county its cars are often crowded with excursionists from San Antonio to various places along the road. Medina Park, not many miles from Castroville and about twenty miles from San Antonio, has become a popular place of amusement; entire families bring tents and camping outfits, and enjoy themselves in a thousand pleasant ways, often remaining several days, and returning to the daily routine of life with the determination to revisit this delightful park. It is situated on a plateau near the Medina River, shaded by huge oaks that are covered with natural swings of mustang vines, so inviting to childish imaginations. A spacious pavilion has been recently erected here by the company, besides many comfortable cottages. Rustic benches and chairs in natural arbors invite to rest, and a band of music frequently enlivens the scene with the bugle-call to the dance.

Along the line new places of future importance are being laid out, in towns such as Lacoste, Summit, Hondo City and D'Hanis. Other places in the county are growing and prospering, as Quihi, New Fountain, Sico, Francisco, Perez and others; all thriving little villages, supplied with churches and having excellent educational advantages.

UVALDE COUNTY.

From Medina the "Sunset" passes into this remarkable county, of which, up to the present time, very little has been written, although it is known to have strong attractions for the health-seeker, tourist, capitalist, laborer and sportsman. A volume might be filled with the beauties, grandeur and wild scenery of the Nueces Mountains alone. This county, formerly known as Rio Brava, the old Spanish name, is situated in southwestern Texas, on the 23d degree of longitude west from Washington, and between the 29th and 30th parallels of north latitude. It contains 1300 square miles, with a population of four thousand, which is constantly increasing, and is composed of Americans and Mexicans, in the proportion of two of the former to one of the latter.

The watercourses all take their rise among the heights in the northern part of the county, and flow in a southerly direction, dividing themselves pretty equally throughout the county. The Sabinal River is one of the chief streams and has fertile land extending for many miles on either side. Its waters are beautiful, clear and cool, and are inhabited by the finest game-fish in the land, which can be caught in great abundance. The attention of the explorer upon entering this rich valley is at once attracted by the numerous and large groves of immense cypress to be found along the river borders. These giants of the forests, judging from their huge proportions, are probably the oldest in the land and the most serviceable for the uses of civilization. One may pass a life time in travel and not find anywhere their equal in size. This river, the Sabinal, is principally formed by Blanchero, Turkey, Canon and An-Flaco Creeks, which, from their sinuous courses, like the river, render irrigation easy, and the raising of crops not a matter of chance, but a certainty.

Sabinal, which takes its name from the river, is a station seventy-one miles from San Antonio on the Sunset Route, and is easily accessible from the farming districts. Its central location and accessibility with regard to the agricultural interests, render its rapid growth and prosperity fixed facts. It is the centre through which all of the exports and imports will pass. The whole length of this river valley, from the northern boundary to the south line of the county, can be rendered, by irrigation, one ever-blooming garden spot. Near the southern border of the county, this stream empties into the Rio Frio, a somewhat larger river and one possessed of the same valuable characteristics, but on a much larger scale. On the latter river is Chatfield, another station, peculiarly fitted for the shipment and handling of stock from the extensive ranges in that section. These two stations place every farmer and stock-raiser in the county in easy and quick connection by rail with the great trade centres of the South and West. Next to the agricultural and pastoral advantages above mentioned, may be noticed the beautiful scenery and landscapes presented on every side to the passenger looking from the window of a parlor car as it glides along over a new and fine steel rail track. These two rivers and the beautiful Rio Leona, with their numerous tributaries, formed from never-failing springs, afford a copious water supply. But to improve on nature and increase this abundance, a strong company has been formed in New York for the purpose of cutting a canal for irrigating purposes, from a point twenty miles northwest of Uvalde, on the Nueces River, to a designated point on the Leona. This canal, while it will cross only about twelve miles of country, will bring about 30,000 acres of valuable land subject to irrigation and increase the certainty of raising good crops. The fall of the country and the inexhaustible supply of water afforded by the Nueces, will make this canal equal to a natural river, rushing through the country and distributing its benefits without stint and alike to all settlers on its borders.

Uvalde, the county seat, a short distance from the railroad and about ninety-three miles from San Antonio, is situated on a romantic lakelet in the southern central portion of the county. It is the largest town in the county, having about 1200 Americans and 300 Mexicans within its corporate limits. This forms the chief point of commercial importance in this section, and promises to reach metropolitan proportions in the course of a few years. A large and commodious hotel has just been fin-

ished for the special accommodation of tourists, invalids and trading men from the interior. Even this with the other hotel accommodations is hardly equal to the demand of the public. There are several wholesale business houses there, doing a business that amounts to $200,000 per annum. Farmers, stockmen and rancheros from all parts of the country deal with merchants here, and through them reach the greater cities. The town is prettily and tastily laid out on the banks of the limpid Leona, which river, by a curious freak, forms a lovely lakelet about a mile from town. Of this we will speak more at length hereafter.

The numerous and thick groves of live-oak, post-oak, hackberry and pecan lend much to the other natural attractions of the place, and

go far, with the extraordinary beauties of the river and its rugged banks, to make a perfect paradise for the invalid and seeker after health and recreation.

This place, being situated on the road that has been for many years past the great highway for traffic between the more thickly settled sections of Texas and northwestern Mexico with its numerous towns, is frequently filled with large wagon trains or long files of pack-mules carrying merchandise and products and is thus made the scene of novel and interesting sights. Its near proximity to the Nueces River forms another point of interest, as it makes practicable a trip through this wild, romantic and unsettled region, where we find the wild turkey, Mexican lion, black-tail deer, mountain sheep occasionally, and even buffalo.

In the Nueces range of mountains, higher up in the county, but not out of reach of this town, are vast mineral resources that some day will be developed and will form a source of untold wealth to this section. There is also an ample supply of timber for domestic and agricultural purposes. The topography of this district is generally broken and diversified, but its elevated nature renders this a most excellent resort for invalids. This dry, bracing air of a thousand feet above the level of the sea is pronounced a panacea for all pulmonary diseases, even though they be in a most advanced stage. Among the very hills nature has placed a wonderfully beautiful lake, filled it with the finest table-fish in the world and bordered it with forest trees of the densest foliage. . It is supplied by innumerable subterranean springs that boil out from the various strata of limestone that form the basin. The depth in places is an hundred feet, and the water is so clear that white rock nearly fifty feet down can be clearly seen. Fish swimming far below the surface are almost as distinctly visible as birds flying in the open air. Years ago Indians of many tribes made this romantic retreat their camping-ground. Here in their wigwams they sharpened the savage tomahawk and whetted the barbarous scalping-knife for bloody butchery, or smoked the pipe of peace as they watched their toiling squaws dress the buffalo robe or tan the hide of the wild deer. Here they taught the dusky youth the skilful use of the bow and arrow, or in the frail bark canoe skimmed the silvery surface of the lake in pursuit of goggle-eyed perch and beautiful mountain trout. This is a spot where life is worth living and where nature has been so lavish in the dispensation of her beauties, that there is companionship, even though the haunts of man be far away. .

During the rainy season, when the Leona River is very high, it is believed to have a subterranean connection with this remarkable basin, and even has been known to overflow the surrounding land and pour in an extra supply of water. Going further along the line of the railroad the Nueces River and bridge are the next points of interest. It is a beautiful stream with characteristics similar to those we have just described. The bridge is solidly set upon stone piers, the material in which was taken from a quarry not far away, and is of the most durable quality. Many rancheros in this section have their houses constructed of this stone. Seen by the traveler the abodes impress him strongly as being the homes of comfort and happiness. Not far distant from Uvalde is Fort Inge, and a pretty village called Waresville, on the banks of the Nueces River.

KINNEY COUNTY.

The road leaving in this vicinity Uvalde County, next enters Kinney, one of the extreme western counties of the state, bordering on the Rio Grande. It is situated between the 23d and 24th degrees of longitude west from Washington, and the 29th and 30th parallels of latitude. It is distant 125 miles west from San Antonio, and has an area of over 1400 square miles, with a population of 2000 Americans and Mexicans.

It is almost exclusively a grazing country, being adapted expecially to sheep husbandry, of which we will speak more fully in time.

With a system of irrigation which is very successfully managed by those who undertake it, much land is cultivated and yields a large profit to the farmer. Crops by this means are kept growing from year in to year out. As a winter resort this county is unsurpassed.

Leaving Uvalde, prepared for several weeks of camp life, we take an ambulance to explore the county more thoroughly in all directions. This trip we will briefly describe in a chapter appropriated to that purpose. At present our writing will be confined in a general way to the counties and territory traversed by the Sunset Route. The main line runs from Uvalde, crossing Nueces River, Turkey, San Moras and San Filipe Creeks to Del Rio, thence to the Rio Grande, along which it passes to the mouth of Devil's River and crosses it a few miles above, entering a very picturesque canon, reaching to

CROCKETT COUNTY,

one of the largest though yet unorganized counties in this section of the state. This county is situated south of the 31st degree of north latitude, and between the 23d and 25th degrees of longitude west from Washington. The Pecos River flows along the western boundary, with many tributary creeks and springs. On the southwest it is bounded by very picturesque scenery, and in the sight of many imposing views of the beautiful State of Coahuila—Mexico.

Generally this vast county, though in many places it approaches the mountainous, is undulating. In the more level parts near the river the fertility of the soil is unsurpassed. Of this 16,000 square miles very little is under cultivation, and timber is somewhat scarce, but for grazing purposes it is not excelled anywhere. Sheep husbandry and the raising of goats are followed with unprecedented success. The altitude is 2000 feet above the level of the gulf, and the climate is delightful, with a dry and healthy atmosphere.

This region and others toward El Paso will soon be better explored, and details will be given in an edition of this work next season, which will also contain glimpses of Mexico, Arizona and southern California.

Before entering our ambulance for the journey overland, we will just glance at two western counties traversed by the Eagle Pass branch of the Sunset Route. One of these is

ZAVALLA.

South of Uvalde and adjacent to it, this county has very much the same topography and general characteristics of soil, climate and elevation. It is well watered by the rivers Leona and Nueces, by the creeks Turkey,

MEDINA PICNIC GROUNDS.

Am. Bank Note Co. N.Y.

Las Minas, Oak and by Forked Lake. The land lying along these streams is very productive and in most cases well timbered. A large portion of the better land is under cultivation and yields abundantly to the hand of the industrious husbandman. The railroad penetrates only a portion of the northwestern part, then passes through

MAVERICK COUNTY

to its Texas terminus. This latter county is a large, fine section of country, situated between the 23d and 24th degrees of north latitude. Its area is 1600 square miles and its population nearly 4000. Eagle Pass and the Rio Grande is the present terminus of the great Sunset Route, which stretches almost entirely across the great territory of Texas. This point is 160 miles west of San Antonio and 376 from Houston, the eastern terminus of the road. Just south of Eagle Pass is Fort Duncan, while across the Rio Grande is Piedras Negras. The agricultural advantages of this county are not so extensive as those of other counties we have passed over, but its pastoral greatness will compare favorably with the best sections of the state. The climate is warm but dry and very healthy. It is genial to the rich pasturage grasses that grow here and keeps them green the entire year. This perpetual verdancy of the grazing is the secret of the success of stock-raising here.

The only river of any note besides the Rio Grande is the Tecasquite, but there are the Las Moras and quite a number of other creeks with rills and branches not a great distance from Eagle Pass.

The geological formations on both sides of the river are soon to be explored, and as remarkable discoveries have already been made, we may be able to give some very interesting data in the account of our future rambles up the Rio Grande to El Paso.

Rambles from Uvalde to the Rio Grande,

AND FROM

DEL RIO ACROSS THE COUNTRY TO EAGLE PASS AND PIEDRAS NEGRAS.

WHEN we first traversed Texas from north to south, having come from the snow-covered northern states, it seemed to us so strange to see great cypresses, live-oaks with tangled vines, Spanish dagger, palmettoes, tall, rustling grasses in all tints of refreshing verdure, with thousands of croaking frogs in the midst of water plants, we were in-toxicated with delight, but became in time accustomed to it and also to her almost never-ending freshness of verdure. Now again traversing Texas from east to west, our en-thusiasm becomes renewed, especially by spending part of the winter and fall in the southwest section of this favored country. There are at that time fewer blossoms, and the weather is somewhat cooler than during spring or summer, but in regard to the agreeable out-door life, the refreshing, balmy air on the high plains, the blue sky, and eternal verdure, the intoxicating delight we felt before returned and increased day by day, and we are now more in love with the "Lone Star State" than ever. These highland regions are bound to be-come the favorite winter resorts in the future. From them you will never be out of sight of mountains, beautiful prairies, valleys, wood-land, creeks and rivers. The longer you re-main the more you are infatuated. From San Antonio the cretaceous formation begins, ex-tending northeast toward Austin and south-west toward the Rio Grande. In Uvalde County it changes its course directly to the west, forming a natural line dividing two dis-tinct regions of geological formations, with an elevation gaining very rapidly. This, so little traveled region, we will traverse from Uvalde to the Pecos River, and from there along the Rio Grande to Eagle Pass and beyond.

One morning, which came sharply over the great plains, sending a thrill of joy through all nature, we are on the way to regions formerly abounding with wild horses, and black with buffalo. Now we meet on the highway men and women vigorous and quick with the glow of health, then a "prairie schooner" crowded with prattling children, extending their saluta-tory, "Howdy!" alike to native or foreigner. Everything is sunshine and exuberant life. Our outfit consists of a well covered four-seated ambulance and two good horses, with an old

stage-driver as pilot, who years ago traversed the plains toward Mexico, El Paso and Santa Fe. Soon we pass a Mexican town, get some glimpses of Uvalde, and we are on our way to Turkey Creek Valley. We pass some fine fields fenced in with stone walls, showing ditches for purposes of irrigation. Fort Inge we leave to the right, and ascending a high plateau we overlook the Nueces valley, with the clear stream, from which its name is taken, gliding through the dense forests toward the Mexican Gulf. Occasionally the silvery sur-face reflecting the bright sunshine peeps up between high and rugged bluffs. In the dis-tance looms up the blue mountain range from whence all these streams are supplied. De-scending, we pass from time to time wagon trains returning from Fort Clark, Laredo or Eagle Pass; finally passing through clumps of live-oak or mesquite we come to the Nueces River, and upon its banks pitch our first camp. It is a rule of importance never to "pitch tent" too near the river, but select a high elevation for protection against a sudden rise in the river of many feet, which is liable to occur without the slighest warning. Another rule is to get beneath shade-trees, to prevent exposure to the heavy dew of the night. Having taken these precautions we are soon settled for the night around a blazing fire.

Not far from us is a camp of hunters just returning from an excursion to the Nueces mountains and canons. Our driver knowing the leader of the party introduces us, and soon an interesting chat is in progress which lasts until long after midnight. A word or two from the experience of these hunters might not be totally without interest to some of our readers of an adventurous turn of mind. They were bountifully provided with ammunition and with good saddle and pack-horses. Their course was due north along mostly the right bank of the Nueces River. About twenty miles up they found plenty of game, and about the break of day brought down from their roosts what turkeys they wanted. Farther up, a black bear, caught in the act of feeding upon the root of bear grass, a favorite food, breathed his last. A Mexican lion prowling around in the evening fell reluctantly at the crack of the unerring rifle. This beast, however, is rather scarce and his skin is prized highly. We made an offer for the one they had, but were refused. Of wolves they found plenty in this locality, but shot only a couple of black coyotes, the pelt of which is also highly prized. All species of wolves are cowardly and will do no more than brave a lamb or some such helpless creature. They make night hideous to campers by gathering round at a safe distance and howling at times in piteous shrieks, then in fierce cries.

This party, venturing out from the river, discovered rich grazing grounds above the high water-mark. They jumped a wild-cat, but not having dogs made no attempt to pursue him. Mule-eared and cotton-tail rabbits darted from "the bush" on every side. Ducks and grouse of limited varieties frequent this region but quail are very abundant. Of the quail, the Mexican, the Massener and the common Texas bird predominate. The chapparal cock, with long tail and brown plumage, and with wonderful fleetness of foot, by which the necessity of taking wing is most always avoided, may be seen by any hunter. The marked peculiarity of this bird is its solitary habits, being rarely ever seen in company with any other. It has no value for the hunter.

In attempting to extend their hunt further, they came to such broken country that a return for rest, the refreshment of the horses and the saving of pelts that had been taken, was absolutely necessary. Among the furs in their possession was that of the black-tail and other deer, the wolf, the mule-eared and cotton-tail rabbit, the bear and the raccoon. The buffalo and antelope are to be found here, but are somewhat rare. The sportsman at Uvalde can equip himself with a complete outfit and have days or weeks of the best sport to be had anywhere in the land. A hunting party has to keep a sharp out-look for Indians, who sometimes waylay to kill and rob. A guide should always be engaged on such extensive trips.

Some specimens of rich ore, iron and copper, were in possession of the party, but it will take a scientific party to investigate and locate the mines, also to estimate their wealth. Whenever this is done a new and great source of wealth will be opened to the public and enterprising capitalists of the world.

After a night of refreshing sleep and a hearty but plain breakfast, we parted with our hunting friends the next morning, and continued our journey toward Turkey Creek. But just before starting we enjoyed the sight of some beautiful landscapes both up and down the stream. Its course is very crooked here, its banks high and rugged, its current swift and clear, its bottom pebbly and many feet from glistening wavelets on the surface. Black bass and blue-cat are the predominating fish and can be caught in abundance. Strata of limestone streaking the high bluffs beautify the landscape very much. Taking the road, we passed during the journey not only much beautiful scenery but rich and productive prairie lands that are sufficiently elevated for the farmer and should be converted into beautiful rural homes.

Before coming to Turkey Creek Springs, but not far from them, are some beautiful groves of pecan, which form the camping-ground for travelers and teamsters, and are the pride of Mr. Cline, their owner, a great hunter and explorer. The springs bubble from the rocky ground and form immediately Turkey Creek, a tributary of the Nueces River. Many of these trees are so close to the banks of the creek that the roots form a perfect net-work, dipping down to the water's edge. The girth of these nut-bearing trees frequently measures eight feet, ranging down to four or five. The branches often reach so far out that they mingle with those coming from the opposite side, and form a protecting green veil over the water that moistens their roots with nourishment as it flows on toward the Nueces. We reached this enchanting spot soon enough to take a short morning stroll and enjoy the sublime scenery of nature, and catch that meditative turn of mind that it so often imparts. Among the branches of these forest trees leaped the cat and fox-squirrel, and were heard the rich notes of the mocking-bird, the shrill voice of the lark and the gentle cooing of the dove. A thousand chirps and twitters came from every side, and the babbling waters played a sweet accompaniment as they toiled on seeking their own level.

Remounting our ambulance we pushed forward and soon arrived at a station composed of several corrals and one store. Here we met a good German, Mr. Cline, of thirty-five years' experience in western Texas. He being a man of culture, shrewdness, and a keen eye for business, erected a large and commodious hotel here, with all modern improvements, to entertain visitors who come here to spend the winter season, with the luxuries of the day. The Sunset Route has made this an important station on account of the superior accommodations of this hotel as a winter resort, and the fitness of Mr. Cline, the proprietor, to take proper care of his guests. Following Turkey Creek, which gets its name from the abundance of wild turkey found along its banks, we advanced toward the Nueces River, where the two form a junction. On this stretch of road any amount of deer, duck, snipe, geese and other game are found. Even the bear hunter finds good game among the thickets and bluffs of this section.

Our driver, who was an old Texan, and had lived in this region many years ago, said that the prairie here and further toward the Rio Grande, in earlier days, was perfectly nude of shrubbery, but for some years past had been growing mesquite rapidly, which aided the ground by its shade to retain moisture, which in turn produced grass abundantly, and is now making a fine grazing country of what was once almost without vegetation. The long curly mesquite grass that grows here now all the year round is happily adapted to the fattening of stock. This new course of nature, it is thought, resulted from the prevention of prairie fires, that once every year used to be started by the Indians, and swept everything in the shape of ligneous material before them. An air line from Uvalde to this point would not be more than twenty miles, but the zigzag course of our road makes it about thirty. Hence we took the road toward Fort Clark and Brackettville, which are not over thirty-five miles distant. This road on either side is bounded by rolling prairies, which sometimes break into ravines and again stretch away, terminating in picturesque hills or ridges. During the dry season these ravines furnish to travelers, hunters and herdsmen their chief supply of water from " water holes " that are formed in their beds. When the dry season is extraordinarily long these sometimes dry up too.

This stretch of country is used for nothing but grazing, the only live water being Elm Creek, about twelve miles distant from Brackettville. On its banks we hurriedly drove, as dark was drawing on rapidly, pitched our camp and slept until two o'clock in the morning, when, on account of the heavy dew, we built a fire. This was a signal for a concert by wolves, which was uninterrupted, except by the hideous notes of a screech owl, until the moon rose, when we gladly struck camp and sought the highway toward Brackettville. On the route we met some suspicious looking characters, but they, mistaking us by the appearance of our ambulance for United States officers, willingly allowed us to pass without interruption. As daylight began to dawn in the east we neared Brackettville, and by sunrise we were in town. Our driver had prepared us for a good breakfast and rest by telling us of a house kept by a former veterinary surgeon, who had allied himself matrimonially with a Mexican senora. Our expectations had not been raised in vain, for we drove up to a beautiful stone residence, surrounded with low-roofed outhouses, and a yard filled with many tropical flowers and plants. The house was built on a plan well suited to the warm climate, having verandas and piazzas on every side. The hail of our driver brought a response from a deep bass voice, the owner of which, upon being informed who we were, extended an invitation to come in.

From the front gallery we were ushered into a neat clean room, in which were two beds, very inviting to the weary traveler, and other furniture betokening refinement and culture. Soon a blazing fire burned in the fire-place and melted the stiffness of our limbs. Taking a good wash and drying our hands and faces on clean white towels, we felt like new men. Breakfast was announced, and we were ushered into a handsome eating-room and beheld a right royal feast, consisting of tenderloin beef and venison steaks, mutton chops, spring chicken, fresh butter, milk and eggs and well prepared birds. This was the work of the lady of the house, assisted by her beautiful daughter, half Mexican and half American, and, sooth to say, it was done to perfection. Not the Menger House of San Antonio, nor even the best hotel in the land, could excel in quality or preparation the meal that was spread before us. It is needless to say that we enjoyed it immensely. Breakfast finished, we looked through the house and found many evidences of taste and culture, which we later learned had been placed there through the instrumentality of the accomplished daughter of our host. From this place, which was on a rather elevated location, could be seen a large portion of the town, with Fort Clark just opposite on the outskirts, and forming an impressive view. Beneath, the Las Moras Springs, with

their surrounding groves; to the west, an extensive prairie, which on its far side was bounded by the Las Moras range of mountains.

On account, probably, of the topographical roughness of the country surrounding, the railroad does not run directly through this place, but passes near enough to make it well worth the time and trouble of the tourist to make a visit here, if for no other purpose than to see Fort Clark. This government post is a thousand feet above the level of the sea, and may be termed the look-out tower of all the surrounding country. The officers stationed here are men of the first rank, and are always cordial and hospitable in their treatment of visitors. In one of their comfortable ambulances a trip to the Las Moras Springs, among their natural and artificial groves, and encircled by elegant drives, is a rare treat.

The landscape scenery in this vicinity would be inspiration to poet or painter, and would challenge comparison with scenery of a similar character anywhere. The Mexican women and girls washing, the children playing around, the water-carts making their constant trips, the mischievous boy ever playing some prank, the soldiers lolling about the springs with a sort of dolce-far-niente air, and the herder resting from his arduous labors—form an animated scene that is probably not to be found elsewhere in this section of country. Far to the west and south valley after valley discloses its charms, and in the dim distance the San Rosa Mountains of Mexico draw their curving profiles upon the blue sky. Before you a sharp contour of the winding course of the Rio Grande is distinctly marked. On the right the Devil's River range of mountains and the Coahouila range attract the eye, and with their beauty invite the traveler, in terms whose silent eloquence is irresistible, to come to this land of promise and of plenty. Acting on this invitation, we renew our course and drive west.

We have now, in an air line, made about forty-five miles, but in our zigzag course made about double that, and have thereby seen more of the country than if we had come over the regularly traveled route. After receiving the cordial treatment and the comforts of Sargent's hotel, we are surprised at having to pay only fifty cents a meal, and take our leave with many regrets. Leaving Fort Clark behind, we cast a glance around us and see the adjoining rocky ridge of limestone, which we ascend starting toward Del Rio. To the south the Fort Clark and Las Morras Hills—the latter with their noted Sentinel Mountain—form a panoramic view of the country that approaches grandeur in its effect.

We reached a high plateau covered thickly with mesquite and chapparal groves, underlaid with rich pasturage grass. The quail, lark, dove and rabbit frequently flee from us as we advance. The Piedras Pinto Mountains lie to the north, and afford a pretty view as we cross the Arroyo Piedras, a clear, swift-running stream bordered with beautiful groves of trees and plots of shrubbery. Ascending to the plateau after leaving the bed of the latter stream, we are shown several ranches, where by attention to the herds large fortunes have been accumulated. Some of these live stock farms look rather dilapidated, but to us each one has a history fraught with deep interest. In the distance the Las Moras Mountains are seen with greater distinctness than ever before by us, and continue to be seen more plainly as we drive forward. Fording the Zoquete Creek, night catches us on Maverick Creek, and we pitch tent. These two streams, including the Sycamore and others, take their rise in the Las Moras Mountains, and flow in the same direction. The landscape here has a peculiar appearance on account of the thickly growing Spanish dagger intermixed with cactus and yucca. Flocks of sheep numbering thousands browse from hill to hill, living upon the luxuriant grass.

While selecting our camping ground a new bird, the mountain falcon or Mexican eagle, is seen; from here on it becomes quite common, while the American eagle is to be seen at intervals. Having selected a spot some distance from the highway, we sleep for the night and resume before daylight our journey toward the Del Rio.

The moon is in its last quarter, and myriads of stars give light sufficient to see the windings of the dim road, until by-and-by roseate Aurora begins to smile from the east and tip with her magic pencil the leaves of the trees and grasses of the fields. It is a gorgeous sight to see the myriads of dewdrops, pendant from as many twigs and grass blades, glistening like diamonds as far as the eye can reach. The lofty bluffs of the Sycamore River are bathed in a flood of sunlight, while the waters below are gilded by scattered beams. Crossing this stream, a drive of several hours over a country similar to what has been described brought us to Filipe Springs, the next point of interest.

Descending a gentle slope, we came to the margin of a clear water basin. It is an immense bowl of cool crystal water, boiling and bubbling as if heated from the furnaces of Hades. Its depth is over sixty feet, and the water is so clear that the strata of limestone are plainly visible at the bottom. Huge trout, fruit for the "lone fisherman," spurt through the clear liquid, and seem at times to rest upon the white rocks. The white limestone at the bottom takes the shape of a beautiful water cave, which is distinctly visible to one peeping from the water's edge, and impresses the mind with its marvelous beauty and unusual appearance. A channel below forms the outlet for this great volume of water, and only a short distance down breaks into a rapid cascade, where the æsthetic eye of the artist may feed with ecstasy and the practical optic of the artisan see visions of great mills and factories in lively operation. Just above the Filipe Springs is another group, which forms a considerable body of water, the surface of which is covered with water-lilies and other aquatic plants. These appear to have direct connection with the others, and there seems to be but little doubt that there is an underground connection between them. To the south is San Filipe military post, which, like Fort Duncan, is one of the eyes of our government looking after invasions and depredations upon its territory. This has natural fortifications, and is a valuable stronghold for the United States troops.

It is anticipated that either the railroad company or private individuals will convert the above natural advantages into a desirable health resort by erecting a handsome and commodious building for the entertainment of invalids, tourists and sportsmen. In addition to the excellent bathing facilities offered there is an abundance of game, a genial climate with healthy atmosphere, and last, but not least, the sublime scenery presented by the far-reaching prairies, the San Rosa Mountains, the Rio Grande Valley, and Devil's River with its precipitous bluffs.

A drive of two miles crossing the upper springs brings us to Del Rio, a place of great importance as a winter resort. In a bee-line this is thirty-five or forty miles from Fort Clark, but our course, as usual, made it much greater. Here we were received at Freeman's Hotel, where the accommodations are good and a warm welcome is extended to travelers like ourselves. We find some members of the

EAGLE PASS—FANDANGO BY MOONLIGHT.

Engineer corps and afterward join their party. Del Rio was, before railroad time, like Uvalde, a place of much importance, the stage-coaches on the old Santa Fe route and other lines coming and going every day. It is a general rendezvous for herders, traders, speculators, rancheros, officers, soldiers and sometimes smugglers. Wholesale houses here formerly got their supplies by wagon from San Antonio, and furnished the whole country for miles around with merchandise and supplies. Volumes might be filled with thrilling adventures and stories of frontier towns, but want of room now forbids what we may say at some future time. The attractions of the town have been for some time past, and are now, constantly growing, and will eventually make this one of the most important towns on the Pacific extension. Lately a few Indian scares have alarmed newcomers, but no attacks have occurred as in years past. The town itself, with a population of about 600 Americans and Mexicans, is regularly and tastily laid out. All through it run irrigation canals, which are managed judiciously by a company for that purpose. This company also has for rent large tracts of land subject to irrigation.

On account of the great water-power of the San Filipe River this place may at some future day become a great manufacturing centre. Along this river are over 4000 acres of irrigated land devoted to the cultivation of corn alone. Here are also many beautiful farms in a high state of cultivation.

Almost within the corporate limits of the town is the celebrated Sugar Loaf Mountain, composed of former river strata, which became by some means isolated. It appears to have been once a part of, and is likely a renewal of, the old chain of hills which bounded this valley, but which have disappeared before the hand of time ; while this, being more durable, is still standing. Its top is about 200 feet above the ordinary level of the ground, and from it a magnificent view is to be had in any direction, and one which for unbroken range of sight is rarely equaled. Standing upon the apex of this cone, at your feet is Del Rio, with its water streets in the form of irrigating ditches. Across the river is the old Mexican town, antiquated in appearance, but none the less beautiful ; this is connected by numerous foot logs or rustic bridges built in a rude manner by the natives. A number of water mills, with their ponderous wheels moving slowly around, lend a pleasant effect to the impression made. For miles along the river banks beautiful farms with their neat cottages—something rare in this section—succeed each other in close proximity. A short distance off is the Rio Grande, and far beyond are the blue ridges of the Santa Rosa and Coahuila ranges of mountains. The Rio Grande, in its serpentine course, doubles and quadruples itself over and over for miles within view. Looking to the east, the military posts are seen, and their embankments bristling with warlike associations.

At the foot of these battlements the river steals along, forming a natural barrier against the invader. The eye reaching still further meets the foot hills of the Las Moras and Nueces ranges, and soon climbs the rock-ribbed sides of the mountain ranges themselves, the latter of which is reputed great in mineral wealth. In another direction, and near by, the custom-house shows up in full view, also the ferry and ford between Mexico and the United States.

This is one of the many desirable places for the invalid to visit. Not only is the country healthy, but it is full of novelties and attractions that will deeply interest his mind and imagination for weeks or months. Sources of inexhaustible wealth, yet untouched, are here for any one who desires to invest capital and reap a rich harvest. Material upon which the scientific mind can feed for a lifetime abounds.

The mountains of Texas form the great sanitariums of nature, and should be made the resort of those persons who, in other states, have contracted diseases that the rigor of the climate will never allow to heal. Many instances of marvelous cures by the climate and mountainous district can be found in the region of San Antonio, where invalids have moved expecting to die, but in many cases entirely recovered their health ; and in others added years of happiness to what otherwise would have been miserable existences ending in early deaths. Some far-sighted capitalist or wealthy philanthropist will no doubt soon erect a commodious structure, somewhere in these sources of health, for the accommodation and use of invalids and seekers after health. Such a building should be in an altitude of pure mountain air, and provided with all appliances conducive to the comfort and protection of the occupants. Ample provision should be made for northers and the deleterious effects of the sudden changes they sometimes produce in the temperature. "Wet northers" last twelve or eighteen hours, and are the most disagreeable. "Dry northers," though cold, are not so disagreeable, and may even be beneficial in purifying the atmosphere. It is seldom that either of them lasts long enough or is sufficiently cold to damage fruit trees or kill stock, though many instances can be found in the history of this country where such has been the case. The dry norther should be regarded more as a messenger of health than otherwise ; for, being free from miasma, its purifying effect is self-evident ; and being fraught with electricity, that great health restorer, it has a very exhilarating and buoyant effect upon the human system.

To resume our journey we crossed west from Del Rio and came upon a rolling prairie covered with mesquite, cactus, some species of aloes and Spanish dagger, intermixed with a fine growth of rich grass. Here we passed a large sheep ranche, with its stables, corrals, pens and lots, where the owner gathered and counted his herds by the thousand. Passing on, the prairie became broken by bluffs and small canons and the roadway almost obliterated by rocks; but we, wishing to reach the camp of May Polk (the head of one of the Engineer corps), situated about eighteen miles away, on the Rio Grande, pushed forward to accomplish the journey. The roughness of the country greatly impeded our progress, and in one instance came near upsetting our vehicle and turning out its contents. We escaped, however, and finally reached the camp of the Engineer corps, where we were cordially welcomed and hospitably treated. During the latter part of this drive darkness approached so rapidly that we had no time for observations; but on awakening from a comfortable rest on the ground-floor next morning ample reward came to us in the sublime scenery that sunrise revealed. Our tents, only a few feet from the bluffs overlooking the Rio Grande, stood beneath a grove of giant pecans. From here we took our first look, and were vividly reminded of the beauties of the Hudson River. The yellow waters stretching toward the orient glistened and sparkled in the morning sunlight like a field of molten gold. In marked contrast to this was the emerald hue of the numberless groves of pecan trees lining its banks as far as the eye could reach. The gray rocks and precipitous canons further increased the varied colors of the sublime landscape and heightened the wonderful effect. Except the Missouri, perhaps, the Rio Grande is different in appearance and characteristics from all rivers in the world. The turbid, boiling, surging water during a freshet, when giant trees are swept down by its current like mere twigs, fills the stoutest heart with terror. Especially is this the case when it is remembered that these rises come down like a great avalanche without any premonitory warning. One moment the country is dry, the next it is the bottom of a living, raving sea that inundates the valley in the twinkling of an eye. But its habitual state is a placid, quiet and gentle one. The greater part of the shores of this mighty stream is not wooded, save where the many canons that widen into valleys bear trees and shrubbery of the most beautiful foliage.

These treeless banks rise perpendicularly from the water's edge to a height of many feet, while from their summits the boundless prairies extend east and west for hundreds of miles, without even a twig to obstruct the sight or break their baldness.

This locality where we now stand, though five hundred miles in the interior or from the mouth of the river, will soon be rendered accessible to the world by the iron girdle of the Sunset extension as it circles the bluffs, rocky peaks and hills in this region of romance.

The railroad measures should here be at once inaugurated, in order to bring within the reach of the traveler the wonderful gifts that nature has so benignantly granted this region, and which are now unobserved save by a few adventurous souls. The glorious mountains, rivers and prairies; the beautiful glens, valleys,

canons, hills and dales, decked in their emerald garments of luxuriant and never-fading foliage, should be placed before the eye of the invalid, the tourist and the man of science. The pure and bracing atmosphere should inflate worn-out lungs and impregnate every particle of diseased bodies. The delicious and cool water should delight the palate and erase the seeds of disease from the skin. The healthy and nutritious vegetation of the fertile soil should strengthen the feeble and invigorate the strong. The wonders of this wild nature should be presented to the over-worked and exhausted brain of the business man, to create within him a stimulus to his jaded imagination and a restoring energy to his dormant fancy. To accomplish all of this the hand of the architect and the artisan must erect in this blooming wilderness, this western El Dorado, a monument to their own skill and the potency of capital.

A hotel palace would be the most appropriate and appreciated design for such a monument; its fame would be borne along the

"Sunset Route" from the Pacific Slope to the Atlantic Seaboard. Invalids and pleasure-seekers, from the remotest part of the Union, would be its constant visitors during the winter season; the one, to gain health and to realize the dream of Ponce de Leon; the other, to find fresh fields and pastures new for enjoyment. Just above the junction of Devil's River with the Rio Grande is the spot where this palatial hotel, this temple to the twin divinities of health and pleasure, should be built. There the grandeur of the well-named Rio Grande would delight the admiring eye from day to day; there the wind-swept bluffs of the Devil's River would become the scene of morning promenades and evening drives for the visitors from every city—not only distant cities, but those contiguous to the line of the railway. Such a structure as this, placed upon one of the elevated plateaus that far overreach the greatest altitude of the high water-mark, would be simply a terrestrial Eden—a paradise whose rock-bound gateway could only be reached by the Sunset Route, and whose portals would open when her engines shrieked for admission. Out of danger of high-water, disease, extreme heat or extreme cold, and free from all the business cares that so poison and distract existence, one could then sit in luxurious apartments, breathing the purest breath of heaven, and beholding on every side the most sublime scenery, perhaps, on earth, for hundreds and hundreds of miles.

This would be happiness, for health and luxury are happiness. At the base of the hotel plateau the broad bosom of the Rio Grande del Norte would heave in violent emotions with its swollen waters, or sigh, as if with the gentlest murmurs of love, from its tranquil depth. Hundreds of feet beneath, its winding course, through the emerald-hued fields, could be followed by the sight, until, like a tiny silver thread, it could still be seen winding its way through a mighty ocean of green. Bounding this broad expanse, the Santa Rosa Mountains of Mexico could be seen lifting their towering tops to the skies.

On the right, a short distance away only, the crystal waters of Devil's River mingle with the saffron waves of the dancing Rio Grande, and for miles and miles they glide along, crystal and saffron, side by side without commingling.

Along the bluffs of these two streams many a spot rich from the lavishness of nature, and many a steep declivity awful from its magnitude, present themselves to the eye. Around the rocky precipices the iron monster winds his way, and, a hundred feet above the terrifying chasms, follows his steel track. Hills and valleys resound with his sharp voice, and every glen, dale and canon echoes his hoarse breathing and ponderous clanking.

The reader, from this attempt to picture the situation, can do no more than get a vague idea of the glorious reality that awaits his coming. Life there becomes a dream, a beautiful vision, a fabric woven from sweets too delightful to seem real.

With all of the pleasures presented by this lovely country, it has a store of wealth that cannot fail to attract the attention of business men. Its exhaustless quarries, its rich iron and copper ores, its fertile alluvial soils, its incalculable water power, are all sources of immense wealth—which are not often found in connection with inexhaustible pleasure and health. The railroad near this point crosses one of the oldest highways known to have existed in Texas. Two hundred years ago it was used by travelers in crossing this immense plain. Not far away are the Plains of the Pecos, where the buffalo may be hunted successfully even at the present day; while in former times they grazed upon the prairies in large herds, and were hunted by the white man in a manner graphically described by Mr. George Catlin in *Catlin's North American*

Indians. We take the liberty of inserting his entertaining account of the hunt:

"Every one of these red sons of the forest (or rather of the prairie) is a knight and lord—his squaws are his slaves. The only things which he deems worthy of his exertions are to mount his snorting steed, with his bow and quiver slung, his arrow-shield upon his arm, and his long lance glistening in the war-parade; or, divested of all his plumes and trappings, armed with a simple bow and quiver, to plunge his steed amongst the flying herds of buffaloes, and with his sinewy bow, which he seldom bends in vain, to drive deep to life's fountain the whizzing arrow.

"The buffalo herds, which graze in almost countless numbers on these beautiful prairies, afford them an abundance of meat; and so much is it preferred to all other that the deer, the elk, and the antelope sport upon the prairies in herds in the greatest security, as the Indians seldom kill them, unless they want their skins for a dress. The buffalo (or more correctly speaking, the bison) is a noble animal that roams over the vast prairies, from the borders of Mexico on the south to Hudson's Bay on the north. Their size is somewhat above that of our common bullock, and their flesh of a delicious flavor, resembling and equaling fat beef. Their flesh, which is easily procured, furnishes the savages of these vast regions the means of a wholesome and good subsistence, and they live almost exclusively upon it, converting the skins, horns, hoofs and bones to the construction of dresses, shields, bows, etc. The buffalo bull is one of the most formidable and frightful looking animals in the world when excited to resistance; his long shaggy main hangs in great profusion over his neck and shoulders, and often extends quite down to the ground. The cow is less in stature and less ferocious, though not much less wild and frightful in her appearance.

"The mode in which these Indians kill this noble animal is spirited and thrilling in the extreme; and I must, in a future epistle, give you a minute account of it. I have almost daily accompanied parties of Indians to see the fun, and have often shared in it myself, but much oftener ran my horse by their sides, to see how the thing was done and study the modes and expressions of these splendid scenes, which I am industriously putting upon the canvas.

"They are all, or nearly so, killed with arrows and the lance, while at full speed; and the reader may easily imagine that these scenes afford the most spirited and picturesque views of the sporting kind that can possibly be seen.

"At present I will give a little sketch of a bit of fun I joined in yesterday with Mr. McKenzie and a number of his men, without the company or aid of Indians.

"I mentioned the other day that McKenzie's table, from day to day, groans under the weight of buffalo tongues and beavers' tails and other luxuries of this western land. He has within his fort a spacious ice-house, in which he preserves his meat fresh for any length of time required, and sometimes, when his larder runs low, he starts out, rallying some five or six of his best hunters (not to hunt, but to 'go for meat'). He leads the party, mounted on his favorite buffalo horse (*i.e.*, the horse amongst his whole group which is best trained to run the buffalo), trailing a light and short gun in his hand—such a one as he can most easily reload whilst his horse is at full speed.

"Such was the condition of the ice-house yesterday morning, which caused these self-catering gentlemen to cast their eyes with a wishful look over the prairies; and such was the plight in which our host took the lead, and I, and then Mons. Chardon and Batiste Defonde and Tullock (who is a trader amongst the Crows, and is here at this time with a large party of that tribe), and there were several others whose names I do not know.

"As we were mounted and ready to start McKenzie called up some four or five of his men and told them to follow immediately on our trail with as many one-horse carts, which they were to harness up, to bring home the meat. 'Ferry them across the river in the scow,' said he, 'and follow our trail through the bottom; you will find us on the plain yonder, between the Rio Grande and Pecos Rivers, with meat enough to load you home. My watch on yonder bluff has just told us, by his signals, that there are cattle a plenty on that spot, and we are going there as fast as possible.' We all crossed the river, and galloped away a couple of miles or so, when we mounted the bluff; and to be sure, as was said, there was in full view of us a fine herd of some four or five hundred buffaloes, perfectly at rest, and in their own estimation (probably) perfectly secure. Some were grazing and others were lying down and sleeping; we advanced within a mile or so of them in full view and came to a halt. Mons. Chardon 'tossed the feather' (a custom always observed to try the course of the wind), and we commenced 'stripping' as it is termed (*i. e.*, every man strips himself and his horse of every extraneous and unnecessary appendage of dress, etc., that might be an incumbrance in running); hats are laid off, and coats and bullet pouches; sleeves are rolled up, a handkerchief tied tightly around the head and another around the waist, cartridges are prepared and placed in the waistcoat pocket, or a half-dozen bullets 'throwed into the mouth,' etc., etc., all of which take up some ten or fifteen minutes, and is not, in appearance or in effect, unlike a council of war. Our leader lays the whole plan of the chase, and preliminaries all fixed, guns charged and ramrods in our hands, we mount and start for the onset. The horses are all trained for this business, and seem to enter into it with as much enthusiasm, and with as restless a spirit, as the riders themselves. While 'stripping' and mounting they exhibit the most restless impatience; and when

'approaching' (which is, all of us abreast, upon a slow walk, and in a straight line towards the herd, until they discover us and run), they all seem to have caught entirely the spirit of the chase, for the laziest nag amongst them prances with an elasticity in his step, champing his bit, his ears erect, his eyes strained out of his head and fixed upon the game before him, whilst he trembles under the saddle of his rider. In this way we carefully and silently marched until within some forty or fifty rods, when the herd discovered us, wheeled and laid their course in a mass. At this instant we started (and all must start, for no one could check the

fury of those steeds at that moment of excitement), and away all sailed, and over the prairie flew, in a cloud of dust which was raised by their trampling hoofs. McKenzie was foremost in the throng, and soon dashed off amidst the dust and was out of sight—he was after the fattest and the fastest. I had discovered a huge bull whose shoulders towered above the whole band, and I picked my way through the crowd to get alongside of him. I went not for 'meat,' but for a trophy; I wanted his head and horns.

"I dashed along through the thundering mass as they swept away over the plain, scarcely able to tell whether I was on a buffalo's back or my horse—hit, and hooked, and jostled about, till at length I found myself alongside of my game, when I gave him a shot as I passed him. I saw guns flash in several directions about me, but I heard them not. Amidst the trampling throng, Mons. Chardon had wounded a stately bull, and at this moment was passing him again with his piece leveled for another shot; they were both at full speed, and I also, within the reach of the muzzle of my gun, when the bull instantly turned, receiving the horse upon his horns, and the ground received poor Chardon, who made a frog's leap of some twenty feet or more over the bull's back, and almost under my horse's heels. I wheeled my horse as soon as possible and rode back, where lay poor Chardon gasping to start his breath again, and within a few paces of nim his huge victim, with his heels high in the air and the horse lying across him. I dismounted instantly, but Chardon was raising himself on his hands, with his eyes and mouth full of dirt, and feeling for his gun, which lay about thirty feet in advance of him. 'Heaven spare you! Are you hurt, Chardon?' 'Hi—hic, hic, hic, hic—no—hic —no, no—I believe not. Oh, this is not much, Monsieur Cataline—this is nothing new ; but this is a d—d hard piece of ground here. Hic —oh !—hic.' At this the poor fellow fainted, but in a few moments arose, picked up his gun, took his horse by the bit, which then opened its eyes, and with a hic and a ugh—UGHK !— sprang upon its feet, shook off the dirt ; and here we were, all upon our legs again, save the bull, whose fate had been more sad than that of either.

"I turned my eyes in the direction where the herd had gone, and our companions in pursuit, and nothing could be seen of them, nor indication, except the cloud of dust which they left behind them. At a little distance on the right, however, I beheld my huge victim endeavoring to make as much headway as he possibly could from his dangerous ground, upon three legs. I galloped off to him, and at my approach he wheeled around and bristled up for battle ; he seemed to know perfectly well that he could not escape from me, and resolved to meet his enemy and death as bravely as possible.

"I found that my shot had entered him a little too far forward, breaking one of his shoulders and lodging in his breast ; and from his very great weight it was impossible for him to make much advance upon me. As I rode up within a few paces of him he would bristle up with fury enough in his looks alone almost to annihilate me, and making one lunge at me, would fall upon his neck and nose, so that I found the sagacity of my horse alone enough to keep me out of reach of danger ; and I drew from my pocket my sketch-book, laid my gun across my lap and commenced taking his likeness. He stood stiffened up and swelling with awful vengeance, which was sublime for a picture, but which he could not vent upon me. I rode around him and sketched him in numerous attitudes. Sometimes he would lie down, and I would then sketch him ; then throw my cap at him, and, rousing him on his legs, rally a new expression and sketch him again. In this way I added to my sketch-book some invaluable sketches of this grim-visaged monster, who knew not that he was standing for his likeness.

"No man on earth can imagine what is the look and expression of such a subject before him as this was. I defy the world to produce another animal that can look so frightful as a huge buffalo bull, when wounded as he was, turned around for battle and swelling with rage ; his eyes bloodshot and his long shaggy mane hanging to the ground—his mouth open and his horrid rage hissing in streams of smoke and blood from his mouth and through his nostrils, as he is bending forward to spring upon his assailant.

"After I had had the requisite time and opportunity for using my pencil, McKenzie and his companions came walking their exhausted horses back from the chase, and in our rear came four or five carts to carry home the meat. The party met from all quarters around me and my buffalo bull, whom I then shot in the head and finished ; and being seated together for a few minutes, each one took a smoke of the pipe and recited his exploits and his ' coups ' or deaths, when all parties had a hearty laugh at me, as a novice, for having aimed at an old bull whose flesh was not suitable for food, and the carts were escorted on the trail to bring away the meat. I rode back with Mr. McKenzie, who pointed out five cows which he had killed, and all of them selected as the fattest and sleekest of the herd. This astonishing feat was all performed within the distance of one mile ; all were killed at full speed, and every one shot through the heart. In the short space of time required for a horse under 'full whip' to run the distance of one mile he had discharged his gun five, and loaded it four times—selected his animals, and killed at every shot ! There were six or eight others killed at the same time, which altogether furnished, as will be seen, abundance of freight for the carts, which returned, as well as several pack-horses, loaded with the choicest parts, which were cut from the animals and the remainder of the carcasses left a prey for the wolves.

"Such is the mode by which white men live in this country, such the way in which they get their food, and such is one of their delightful amusements—at the hazard of every bone in one's body, to feel the fine and thrilling exhilaration of the chase for a moment, and then as often to upbraid and blame himself for his folly and imprudence."

Our party, after spending several days delightfully in this region, photographing and sketching many of the beautiful landscapes, procured a guide and made all necessary arrangements for a trip to Eagle Pass. The warm hospitality of the chief engineer and his able assistants, whose guests we had been, made us loath to say goodbye, or rather "au revoir," for we may see them again before our next issue, in which we will give a fuller account of the Devil's River and the country surrounding it. We will also give a complete, and we trust an *interesting*, account of the country lying between the mouth of the stream and El Paso.

Having made our preparations, we set out on the old Santa Fe trail, but had not gone far before we saw the necessity of having a guide, as we had driven entirely out of the road, and so were obliged to wait for him to direct us before we could either pursue our way or turn back. On this route we traversed a pastoral country rich beyond description, driving often along the banks of the Rio Grande, and then varying our journey by leaving the river for many miles. We were told that this country was, many years ago, the favorite resort of wild horses ; and as these noble animals will

PAINTED CAVES, RIO GRANDE RIVER.

eventually be exterminated or driven from the state, and never be seen by our readers, we fancy that a description of them, as given by Mr. George Catlin, would not be altogether without interest in these pages:

"The tract of country over which we passed—the upper Rio Grande—is stocked, not only with buffaloes, but with numerous bands of wild horses, many of which we saw every day. There is no other animal on the prairies so wild and so sagacious as the horse, and none other so difficult to come up with. So remarkably keen is their eye that they will generally run 'at the sight' when they are a mile distant, being, no doubt, able to distinguish the character of the enemy that is approaching when at that distance, and when in motion will seldom stop short of three or four miles. I made many attempts to approach them by stealth, when they were grazing or playing their gambols, without ever having been more than once able to succeed. In this instance I left my horse, and, with my friend Chadwick, skulked through a ravine for a couple of miles, until we were at length brought within gunshot of a fine herd of them, when I used my pencil for some time while we were under cover of a little hedge of bushes which effectually screened us from their view.` In this herd we saw all the colors, nearly, that can be seen in a kennel of English hounds. Some were milk-white, some jet-black; others were sorrel and bay and cream color, many were of an iron-gray; and others were pied, containing a variety of colors on the same animal. Their manes were very profuse and hanging in the wildest confusion over their necks and faces, and their long tails swept the ground.

"After we had satisfied our curiosity in looking at these proud and playful animals, we agreed that we would try the experiment of 'creasing' one, as it is termed in this country, which is done by shooting them through the gristle on the top of the neck, which stuns them so that they fall, and are secured with the hobbles on the feet; after which they rise again without fatal injury, This is a practice often resorted to by expert hunters, with good rifles, who are not able to take them in any other way. My friend Joe and I were armed on this occasion each with a light fowling-piece, which has not quite the preciseness in throwing a bullet that a rifle has; and having both leveled our pieces at the withers of a noble, fine-looking iron-gray, we pulled trigger and the poor creature fell, and the rest of the herd were out of sight in a moment. We advanced speedily to him, and had the most inexpressible mortification of finding that we never had thought of hobbles or halters to secure him; and in a few moments more had the still greater mortification, and even anguish, to find that one of our shots had broken the poor creature's neck, and that he was quite dead. The laments of poor Chadwick for the wicked folly of destroying this noble animal were such as I never shall forget; and so guilty did we feel that we agreed that when we joined the regiment we should boast of all the rest of our hunting feats, but never make mention of this.

"The usual mode of taking the wild horses is by throwing the lasso whilst pursuing them at full speed, and dropping a noose over their necks, by which their speed is soon checked, and they are 'choked down.' The lasso is a thong of rawhide some ten or fifteen yards in length, twisted or braided, with a noose fixed at the end of it, which, when the coil of the lasso is thrown out, drops with great certainty over the neck of the animal, which is soon conquered.

"The Indian, when he starts for a wild horse, mounts one of the fleetest he can get, and coiling his lasso on his arm starts off under the 'full whip' till he can enter the band, when he soon gets it over the neck of one of the number; when he instantly dismounts, leaving his own horse, and runs as fast as he can, letting the lasso pass out gradually and carefully through his hands, until the horse falls for want of breath and lies helpless on the ground; at which time the Indian advances slowly towards the horse's head, keeping his lasso tight upon its neck until he fastens a pair of hobbles on the animal's two forefeet, and also loosens the lasso (giving the horse chance to breathe), and gives it a noose around the under jaw, by which he gets great power over the affrighted animal, which is rearing and plunging when it gets breath; and by which, as he advances hand over hand towards the horse's nose, he is able to hold it down and prevent it from throwing itself over on its back, at the hazard of its limbs. By this means he gradually advances until he is able to place his hand on the animal's nose and over its eyes, and at length to breathe in its nostrils, when it soon becomes docile and conquered; so that he has little else to do than to remove the hobbles from its feet and lead or ride it into camp.

"This 'breaking-down' or taming, however, is not without the most desperate trial on the part of the horse, which rears and plunges in every possible way to effect its escape, until its power is exhausted and it becomes covered with foam, and at last yields to the power of man and becomes his willing slave for the rest of its life. By this very rigid treatment the poor animal seems to be so completely conquered that it makes no further struggle for its freedom, but submits quietly ever after, and is led or rode away with very little difficulty. Great care is taken, however, in this and in subsequent treatment not to subdue the spirit of the animal, which is carefully preserved and kept up, although they use them with great severity, being, generally speaking, cruel masters.

"The wild horse of these regions is a small but very powerful animal, with an exceedingly prominent eye, sharp nose, high nostril, small feet and delicate leg, and undoubtedly have sprung from a stock introduced by the Spaniards at the time of the invasion of Mexico, which, having strayed off upon the prairies, have run wild and stocked the plains from this to Lake Winnepeg, two or three thousand miles to the north.

"This useful animal has been of great service to the Indians living on these vast plains, enabling them to take their game more easily, to carry their burdens, etc., and no doubt renders them better and handier service than if they were of a larger and heavier breed. Vast numbers of them are also killed for food by the Indians at seasons when buffaloes and other game are scarce. They subsist themselves both in winter and summer by biting at the grass, which they can always get in sufficient quantities for their food.

"Whilst on our march we met with many droves of these beautiful animals, and several times had the opportunity of seeing the Indians pursue them and take them with the lasso. The first successful instance of the kind was effected by one of our guides and hunters, by the name of Beatte, a Frenchman, whose parents had lived nearly their whole lives in the Osage Village, and who himself had been reared from infancy amongst them; and in a continual life of Indian modes and amusements had acquired all the skill and tact of his Indian teachers, and probably a little more, for he is reputed, without exception, the best hunter in these western regions.

"This instance took place one day whilst the regiment was at its usual halt of an hour, in the middle of the day.

"When the bugle sounded for a halt, and all were dismounted, Beatte and several others of the hunters asked permission of Colonel Dodge to pursue a drove of horses which were then in sight at a distance of a mile or more from us. The permission was given, and they started off, and by following a ravine approached near to the unsuspecting animals, when they broke upon them and pursued them for several miles in full view of the regiment. Several of us had good glasses, with which we could see every movement and every manœuvre. After a race of two or three miles, Beatte was seen with his wild horse down, and the band and the other hunters rapidly leaving him.

"Seeing him in this condition, I galloped off to him as rapidly as possible, and had the satisfaction of seeing the whole operation of 'breaking down' and bringing in the wild animal. When he had conquered the horse in this way, his brother, who was one of the unsuccessful ones in the chase, came riding back and leading up the horse of Beatte, which he had left behind; and after staying with us a few minutes assisted Beatte in leading his conquered wild horse towards the regiment, where it was satisfactorily examined and commented upon, as it was trembling and covered with white foam, until the bugle sounded the signal for marching, when all mounted, and with the rest Beatte astride of his wild horse, which had a buffalo skin girted on its back, and a halter, with a cruel noose around

the under jaw. In this manner the command resumed its march, and Beatte astride of his wild horse, on which he rode quietly and without difficulty until night; the whole thing, the capture, and breaking, all having been accomplished within the space of one hour, our usual and daily halt at midday.

"Several others of these animals were caught in a similar manner during our march by others of our hunters, affording us satisfactory instances of this most extraordinary and almost unaccountable feat.

"The horses that were caught were by no means very valuable specimens, being rather of an ordinary quality; and I saw, to my perfect satisfaction, that the finest of these droves can never be obtained in this way, as they take the lead at once when they are pursued, and in a few moments will be seen half a mile or more ahead of the bulk of the drove which they are leading off. There is not a doubt but there are many very fine and valuable horses

amongst these herds; but it is impossible for the Indian or other hunter to take them, unless it is done by 'creasing' them, as I have before described, which is often done, but always destroys the spirit and character of the animal."

As we drove along our way our reveries of wild horses, and the time when they trampled the very spot, perchance, where we were driving, were quite entertaining. Soon we crossed Filipe Creek and drove upon a table-land that stretched far ahead of us. Its surface was broken considerably by small canons extending out from the Rio Grande, but the soil was fertile. From this we descended into the valley of Sycamore Creek, a lovely and productive bottom. It was well timbered with large pecans and sycamores. This creek we mentioned before as we traveled to Devil's River.

We thought of camping beneath these large trees, but when the guide called our attention to the high-water-marks among the limbs of the trees we pushed ahead and passed some fine flocks of sheep herded by Mexicans.

Large numbers of cattle also grazed upon these prairies, and among them were some of the fattest and finest we had ever seen, of the pure Texas beef. They had never been fed, nor had they ever eaten a particle of cultivated vegetation in their lives, yet they were fat and apparently in the best condition. They were not sheltered by sheds, as they received all the needed protection in severe weather from the heavy timber of the bottom-lands. An owner of some of the cattle, upon seeing our astonishment at the fact that they were Texas stock and yet so fine, explained that all large stock-raisers exchanged bull calves every year, in order to sustain the mixture of blood and prevent the enfeebling effects of "breeding in."

We pitched our camp on a high bank just out of the valley, and were fortunate in obtaining a view of a most magnificent sunset. The sunsets in this portion of the country are phenomenally splendid, especially at this season of the year. In the distance were observable the deep blue outlines of the Santa Rosa Mountains, contrasting exquisitely with the golden splendor of the clouds that had gathered, as has been their invariable custom ever since the world was made, around the sinking orb of day, to pay their last respects to the departing god and usher in with purple pomp the stars of the summer night; it was truly a lovely and divine picture. The next morning, arising from a sweet and undisturbed repose, we beheld even a grander sight, if possible, in the rising of the cloud-attended luminary. The wonders of the west were rivaled by the splendors of the east; each day seemed to us more beautiful than any we had ever seen. We continued our journey with practical eyes, however, and saw on all sides points of interest to the farmer and capitalist, as well as to the artist and poet. The beautiful country we left behind us was equaled by the rich tracts we continued to traverse. We finally crossed several pretty creeks and came back to the Las Moras, which we had crossed higher up on our westward trip. This placed us again in Maverick County; and we cannot find a more suitable time in which to give the story of the origin of the name "Maverick," as used by Texans:

"A certain well known 'colonel' of the name bought an island in one of the rivers and stocked it with a few cattle, proposing to keep his animals where he could find them when he wanted beef or hides. Business entanglements claimed the worthy colonel's attention, and in course of time he well-nigh forgot his island colony. Rounders began to find among their herds ancient bulls and cows, all guiltless of an owner's mark; they came to be counted by thousands, and it was finally discovered that they were runaways from Colonel Maverick's island. The old colonel was informed by herders of his good luck, and told, among other things, that some two thousand bulls were subject to his orders. The last thing recorded in connection with this legend is the colonel's excited speech upon the occasion : 'For Heaven's sake, boys, go and help yourselves!' Thereafter any animal found without a brand was called a 'Maverick,' and duly stamped with the finder's mark."

Between Las Moras and Elm Creek the country is broken, though not well watered. In the distance of sixty miles, between Devil's River and Elm, we saw only about four ranches, although the prairie was covered with flocks.

Looking to the northeast, from whence we came, the country had the appearance of being terraced, it being the decline from the elevated section we had visited. The formations, as we advance, are of limestone and sandstone. The plateau about Olmus Creek is almost level; it is covered with good grasses and rests upon beds of coal. Iron ores are also found there with 50 per cent. clear metal. On Olmus Creek are some interesting specimens of posidonomya, ooliths, also sauroliths and other petrifactions in the has formation.

Leaving this creek, Eagle Pass and Fort Duncan soon appear in sight. The bottom lands of the Rio Grande along here are exceedingly fertile, sugar-cane, corn and cotton growing in the greatest luxuriance; some marshy lands covered with a tall grass are found on the route. Passing through a mesquite thicket we were rejoiced at finding ourselves in Eagle Pass, of which place we will speak at length in our next issue of this work.

For the benefit of many of our readers, we will content ourselves with saying here that the broad valley of the Rio Grande is the same in soil as the valleys of the Colorado and Brazos; it bears the rankest growth of mountain cypress, oak, cedar, hickory, pecan, elm, ash, sycamore, cottonwood and many other trees of noble size. Grapes and other fruit grow and flourish here also.

CROCKETT COUNTY.

The railroad crosses the Devil's River, which is spanned by a beautiful iron bridge high above water-mark, connecting in this way the shores close by the fording—a former old Indian trail, and for years the military highway. The Dom Pedro, or Devil's River, with its emerald-like waters, runs many miles clear and swift in its stony bed, lined on both sides with rich groves. This part of the country has not been organized as yet, and but a limited portion is settled. The county has an area of about 16,000 square miles, of which but very little is under cultivation.

Herders and Indians state, that along Devil's River and its tributaries, along Ricardo Creek up to Willow Spring, Kuechler's Lake, Beaver Lake, Pecan Spring and Camp Hudson Spring, beautiful valleys are found which are well adapted for gardening, ranking as high as any soil of this description in this country.

Crossing the river over the beautiful bridge, we enter Painted Cave Canon, winding our way through rocky walls, a chaos of stones.

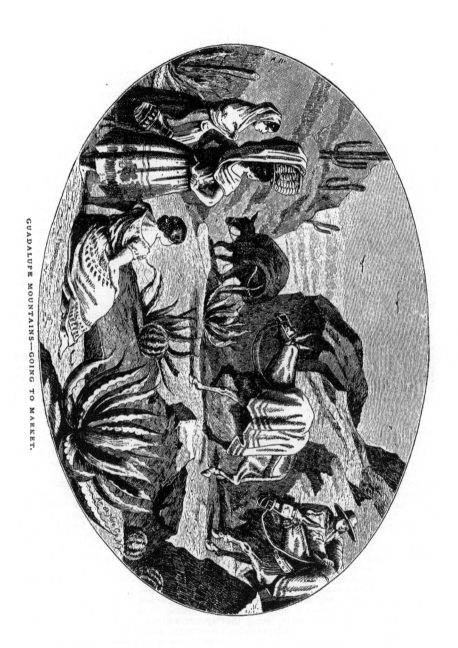

GUADALUPE MOUNTAINS—GOING TO MARKET.

In regard to vegetation, we notice little but cactuses and aloes of various kinds. We will state here, too, that as to engineering, in no region in the United States is it surpassed by any other railroad. The tourist will see canons of the wildest character, and will have a sight of rare grandeur. It will certainly pay him to visit the celebrated "Painted Cave" with its magic chambers. Most of them have been used by those notorious tribes, the Apaches and Comanches, as their hiding-places, and many peculiar figures are either painted or chiseled into its rocky walls.

Everything seems to have been comfortably arranged by nature to invite the tourist to stay and rest. A cool, refreshing spring is close by for the weary teamster, and many a name on the surrounding rocks in the cave is proof and evidence enough that for years these caves have been, and are still, the abode and camping-ground of many travelers, who, further west, may have perished for want of drinkable water.

The western limit of Crockett County is the Pecos River, but before we come to it we have to pass three tunnels through gray limestone mountains, which are respectively twelve, fifteen and sixteen hundred feet in length. The course of the Pecos River is marked on both sides by rich tracts of pasture-land ; the ground is fertile, bare of any growth of brush or timber, but here and there alongside the bluffs there are groves of tall cedars.

Traversing this region on horse-back, there is nothing to indicate the proximity of so large a stream. Descending into a wide canon, an extensive valley smooth with splendid grasses, we stand at once on the banks of the Pecos. Looking down we observe its waters moving on silently, running in a smooth channel. The water is of a dark hue, brackish, and has a decided sulphur and alkaline taste. The current is rather swift, about seven to eight miles an hour ; its depth in many places is twenty feet, and over 100 feet in width. In regard to its bed, it is remarkable for the great number of sharp bends, probably more so than that of any other river on the globe. Many of its side valleys are still a "terra incognita," abounding in wild game ; but what is known of them and explored is easy to irrigate on account of the rapid inclination of this county, about ten feet to the mile.

Some old Mexican settlers in this region, who live along the tributaries farther up the Pecos River, are growing fine corn and wheat crops, especially along the headwaters of Howard Creek. They also tell that elks, panthers, buffaloes, mountain lions, deer, wild turkeys, etc., are found in abundance in the northwestern portion of Crockett County, along the Pecos River, and that there are valleys enough, principally adapted for grazing.

We cross the Pecos River on a bridge of rare excellence, and enter a still larger unorganized county, viz.:

About fifteen miles from the river we come to "Painted Rock Creek." Nature has here painted these rocks with similar colors as those world-renowned "Painted Rocks" on Lake Superior. The creek below is formed by some fine springs, known as "Painted Rock Springs."

Going further down towards the Rio Grande, we traverse a high plateau, passing through wild and picturesque canons, with the Apache Mountains in view, in the direction of Fort Stockton, where we will find nothing but sage, cactus, very seldom water and very little grass. Comanche Creek, near Fort Stockton, in this region makes an exception, inasmuch as it is proved that this whole tract would become fertile by irrigation, for the soil, watered by Comanche Creek, produces the finest kinds of corn, wheat and vegetables. The spring of this creek is said to be one of the most remarkable natural artesian wells in western Texas. Although an immense body of water, the thirsty earth soon absorbs its lively and swift-running waters. Those who intend to visit the Apache Mountains, which are exceedingly rough and abound in chasms, canons, and wild game, will come into an almost unknown region, of which scarcely the outer edges are explored.

As far as it is known, granite and basaltic formations predominate, while along the Pecos River lime and sandstone are found. As to minerals, all kinds of fables and stories exist of gold and silver ores in the Apache regions.

Turning south toward the Rio Grande, the Horsehead Hills come in view, which are likewise very little known, but there are still some first-class ranges for agricultural purposes, and beautiful, fertile valleys with a scenery second to none in regard to sublime grandeur, even to those in the Rocky Mountains.

The elevation in average of this region is a little over 5000 feet, and here you can see the highest peaks of the United States east of the Rocky Mountains.

Pecos County, with about 18,000 square miles, of which the greater portion is still unoccupied, has some large stock ranches and farms. Imagine only that these three immense counties, viz., Pecos, Presidio and El Paso, embrace an area of over 54,000 square miles, which is about the size of New Hampshire, Vermont, Massachusetts, Rhode Island and Connecticut, if put together.

This region is traversed by well-defined mountain ranges, as the Guadalupe, Sacramento, Organ and Chinati Mountains.

Traveling through this portion of the country, one will see rich valleys with fine, nutritious grass, many streams, water-holes, etc., and this it is that makes it so valuable as a grazing country on this continent. Also, along the Rio Grande, the Pecos, the Toyah Rivers and other tributaries, we come to fertile valleys and mountains, the latter being partly covered with a rich growth of large timber, consisting of pine, mountain juniper and some species of oak.

Without doubt there are precious metals yet to be brought to light, and since the railroad has traversed these territories expeditions have been organized to explore these parts of the country ; and by next season valuable results will be gained and published.

The Southern Pacific Route now opens this vast territory, which is rich in minerals, and has a health-giving and invigorating climate. The railroad crosses diagonally the southern portion of Pecos County, and enters,

PRESIDIO COUNTY,

another grand, but still unorganized, country. It is bounded by the Rio Grande with its many tributaries.

Along these valleys fine settlements are found, and capitalists will have good chances for investment. The tourist may rest and lie over at Fort Davis, near Limpia River, about 480 miles from San Antonio, and a little over 220 miles from El Paso ; its elevation above the gulf is about 4700 feet.

This important post was established in 1854, has a charming and healthy climate, and is situated at the outlet of a canon which is here over four hundred feet wide, but is finally lost in the mountains. The country looks sterile, with the exception of some live-oak and cotton-wood clustering along the banks of the Limpia River. In a line north and south across the entrance to the canon are the quarters of the officers, barracks, corrals, the executive office, the guard-house, and stables to accommodate about five hundred horses. The store-houses, the hospital and all those buildings are mostly built of adobe, and form a town by itself. A trip from this post to Wild Rose Pass, and from there to Seven Springs, in the Davis Mountains, will certainly pay the tourist if he wants to see grand scenery. Game is in abundance, and the air is fresh and balmy.

The southern part of Presidio County is yet a " terra incognita." The Rio Grande makes there a great bend, guarded by several military sub-posts. Before long this beautiful spot will be opened to all by a railroad which will be built by the Mexicans along the Rio Grande up to El Paso.

EL PASO COUNTY,

which is easily reached from Presidio, was first settled by Jesuits in 1620. El Paso Valley has an extension of over 140 miles and an average width of about six miles ; its soil is alluvial, rich and productive. The Rio Grande bounds the county on the west and south for a distance of over 100 miles ; on the southeast the boundaries are Presidio and Pecos Counties ; the northern boundary, the 32d degree of north latitude, and divides it from New Mexico. The elevation of this county is about 4000 feet above the gulf ; the whole country is broken and mountainous. The climate is dry and delightful, the rainfall slight, and fine products are raised by irrigation. The population chiefly consists of Mexicans, numbering about 4000, all of whom have their settlements along the Rio Grande Valley ; and now already, through the influence of the railroad, great changes have taken place.

The railroad passes by Fort Quitman, situated several hundred yards east of the Rio Grande and 619 miles from San Antonio. The country around is a sterile sand-prairie, covered here and there with cactuses and mesquite wood. Not far from here the Rocky Mountains rise abruptly and bare of any vegetation. At this post there is no chance to cultivate any vegetables. The buildings are made of adobe, and the fort is supplied by Mexican towns, viz.: Guadalupe, San Ignatio, San Elizario and El Paso.

Farther up, and about three miles from El Paso, there is Fort Bliss, another important military post, situated on Concordia Ranch, about 3600 feet above the level of the gulf. The accommodations at this locality are excellent, and the people live comfortably, excepting the occasional experience of malarial influences from the bottom-lands of the Rio Grande. The portion along the river may be called a real garden spot, especially near El Paso, a city of importance and a railroad terminus. Wheat, corn, rye, barley, onions (the finest in the world), and vegetables of all sorts yield here astonishing crops. Peaches, pears and grapes are cultivated with profit, and in regard to the latter, even European varieties grow very finely, and before long this will rank first as a wine-cultivating country. In general these regions are treeless, except along the streams cottonwood, willow, wild cherry, elm and mesquite are found. In some of the mountain ranges are large cedar forests, many trees over sixty feet in height, rarely met with elsewhere ; and with regard to mineral wealth, important discoveries are on record.

We can say much of frontier life, and will take particular pains to explore those regions, so little known, and of which no correct maps have been published as yet ; and therefore in our next revised issue we shall bring new sketches and a more detailed description of stations and localities of importance, especially to the capitalist and the rancher. We have also in store some very romantic legends of frontier life.

In conclusion, we may say again, this road completed—now ready for the iron horse from New Orleans to San Francisco—is one of the longest, most charming, most varied and delightful in the world.

The great Southwest is now open, and the country will soon be the home of thousands of industrious people.

ADIOS.

LIST OF ILLUSTRATIONS.

THE "STAR and CRESCENT" and "SUNSET" ROUTE:

THE SHORT LINE

And Great National Thoroughfare

— TO —

TEXAS, MEXICO AND THE PACIFIC COAST.

.

Lightning Source UK Ltd.
Milton Keynes UK
UKHW02f2007200818
327529UK00009B/286/P